Lost Civilisations of the Stone Age

Richard Rudgley was born in Hampshire in 1961. After receiving a degree in social antrhropology and religious studies at the School of Oriental and African Studies, University of London, he continued his studies at he Institute of Social and Cultural Anthropology, University of Oxford, in ethnology, museum ethnography and prehistory. He is currently undertaking research into the prehistoric and ancient use of psychoactive plants. In 1991 he became the first winner of theBritish Museum Promcthcus Award which resulted in the publication of his critically acclaimed book *The Alchemy of Culture: Intoxicants in Society* (British Museum Press, 1993). He is also the author of *The Encyclopaedia of Psychoactive Substances* (Little Brown, 1998), *Wildest Dreams: An Anthology of Drug Related Literature* (Little Brown, 1999) and *Secrets of the Stone Age* (Century, 2000), now a major television series. His books have been translated into seven languages. He is married with a daughter and a son and divides his time between Notting Hill and Oxford.

Lost Civilisations of the Stone Age

Richard Rudgley

ARROW

This edition published by Arrow Books Limited 1999

2 4 6 8 10 9 7 5 3

Copyright © Richard Rudgley 1998

Richard Rudgley has asserted his right under the Copyright,
Designs and Patents Act, 1988, to be identified as the author of this work

First published in the United Kingdom in 1998 by Century

Arrow Books Limited
The Random House Group Limited,
20 Vauxhall Bridge Road,
London, SW1V 2SA

Random House Australia (Pty) Limited
20 Alfred Street, Milsons Point, Sydney,
New South Wales 2061, Australia

Random House New Zealand Limited
18 Poland Road, Glenfield
Auckland 10, New Zealand

Random House (Pty) Limited
Endulini, 5 a Jubilee Road,
Parktown 2193, South Africa

The Random House Group Limited Reg. No. 954009

www.randomhouse.co.uk

A CIP catalogue record for this book
is available from the British Library

Papers used by Random House are natural, recyclable
products made from wood grown in sustainable forests.
The manufacturing processes conform to the environmental
regulations of the country of origin

ISBN 0 09 922372 4

Typeset by Deltatype Ltd, Birkenhead, Merseyside
Printed and bound in Norway by
AIT Trondheim AS

To Michael

List of Plates

Acknowledgements

I would like to thank Dr Derek Roe, Director of the Donald Baden-Powell Quaternary Research Centre, University of Oxford, and Dr Andrew Sherratt of the Ashmolean Museum, University of Oxford, for introducing me to the most fascinating of all eras of human existence, namely the Stone Age. Neither of them should be deemed responsible for the views expressed in this book. I would also like to thank Dr Simon Kaner for his help concerning details of recent discoveries at the Sannai-maruyama site in Japan; my friends Michael Carmichael, Virginia Ross and Graham Thomas for their constant enthusiasm and moral support; my agent Andrew Lownie, and my editors Mark Booth and Bruce Nichols. I would also like to acknowledge the indirect influence of Paul Bahn, Robert Bednarik, the late Marija Gimbutas, Alexander Marshack and Colin Renfrew, all of whom have, through their work, illuminated our understanding of the prehistoric world. The love and support of Robin, Rebecca and Benedict is, as always, my foundation stone.

Contents

Introduction

Savage Civilisation

A near-universal theme in the mythologies of the world is that the present state of the world, and more specifically the social world, is in decline – a fall from the Garden of Eden or from a Golden Age. Modern civilisation has turned these traditional mythological assumptions on their head and written a new script, one based on the idea of social progress and evolution. In this new mythology the notion of civilisation (as it is generally understood) replaces Eden and this novel paradise exists not at the beginning of time but, if not right now, then just around the corner. Civilisation is, in the plot of this new mythology, envisaged as a great success story – from prehistoric rags to civilised riches – and it is presented as the final flowering of human achievement born out of a long and interminable struggle against the powers of darkness and ignorance that are represented by the Stone Age.

The way in which the human story has been written to date is so abridged and poorly edited that it has provided us with an account of ourselves which leaves out most of the contents of the early chapters. Despite the fact that prehistory makes up more than 95 per cent of our time on this planet, history, the remaining 5 per cent, makes up at least 95 per cent of most accounts of the human story. The prehistory of humankind is no mere prelude to history; history is rather a colourful and eventful afterword to the Stone Age. In this book I will show how rich and eventful were the contents of these early chapters of the life of our species; how great is the debt of historical societies to their prehistoric counterparts in all spheres of cultural life; and how civilised in many respects were those human cultures that have been reviled as savage. Before doing so I shall show how savage the so-called civilised peoples can be, and how the barbarism of our own culture is projected outward into the geographically remote (modern tribal cultures) as well as into the temporally remote prehistoric cultures.

One might expect that anthropologists, as the representatives of civilised scientific practices in the investigation of tribal societies, would have had a greater respect for their subjects than other colonial groups who had direct experience of the 'natives'. But here we find a sinister skeleton in the cupboard. In 1863 the Anthropological Society of London was set up and included among its members Sir James Hunt, the famous explorer Richard Burton and Robert Knox the anatomist. Burton described it as 'a refuge for Destitute Truth', a reference to the fact that it was the only outlet for his ethnological writings on sexual matters and related subjects that were strictly taboo among the mainstream Victorian intelligentsia. Hunt used the society as a vehicle to express his racist assumptions of the biological basis of white supremacy, juggling anthropometric measurements to lend scientific weight to his prejudices. Within the ranks of the society was an unofficial and informal circle known as the Cannibal Club, with Hunt in the chair calling his comrades to order with a gavel fashioned in the form of a negro's head. Knox was later to resign from his teaching position at Edinburgh University when he was implicated in the notorious criminal activities of Burke and Hare. These partners in crime graduated from grave-robbing to murder in their attempts to keep up with the medical demand for human corpses for dissection. Knox had unwittingly received some of Burke and Hare's unfortunate victims on to his own dissecting table. Despite the scandal surrounding Knox's name, other individuals with anthropological interests did not even bother to obtain their human subjects through middle men but did the dirty work themselves; some even took a certain relish in indulging in this grisly pastime.

The anthropologist James Urry has collected a whole host of grim tales of early anthropologists who behaved no better then the necromancers of the Middle Ages in their respect for the dead. He cites the case of the Russian explorer and ethnographer Nikolai Miklouho-Maclay, who was pursuing science of a most dubious kind in the coastal regions of New Guinea in 1871. He was assisted in his research by a Swedish sailor named Will Olsson and also had a young male Polynesian servant, simply called 'Boy' in his writings. When the boy died from malaria, Miklouho-Maclay was anxious to dispose of the body, fearing that the locals might think he had been personally responsible. Before dumping the corpse in the sea, he was determined to preserve what he could for science. This sordid act is described in its perpetrator's own account, quoted by Urry:

In thinking of the best way to perform the operation, I discovered, to my chagrin, that I did not have a vessel large enough to contain a whole brain. Expecting the natives to appear every hour, most likely with grave intentions, I gave up, not without regret, the idea of preserving the Polynesian's brain but not the chance to obtain a preparation of the larynx with all the muscles, the tongue, etc., as I had promised my former teacher, Professor H. now living in Strasbourg, the larynx of a dark man with all the muscles. Preparing anatomical instruments and a jug with spirit, I returned to Boy's room and cut out the larynx with the tongue and all the muscles. A bit of skin from the forehead and head with hair went into my collection. Olsson, shaking with his fear of the dead man, was holding a candle and Boy's head. As I was cutting the *plexus brachialis*, Boy's hand made a small movement and Olsson, mortally afraid that I was cutting a man still alive, dropped the candle, and we were left in darkness.

Such was the callous nature of the operator that whilst sailing out in his boat to dump the corpse of 'Boy', he was so distracted by the marine life that he went into a sort of scientific reverie deep enough for him to temporarily forget that the corpse was on board. Having surreptitiously and successfully thrown 'Boy' overboard, and satisfied that the sharks would do the rest, he returned to shore to relax over a cup of tea.

A comparable example of total disregard for the remains of native people, this time from South America, has been brought to light by the Oxford anthropologist Peter Rivière. It concerns the activities of a German traveller in British Guiana in the 1840s named Richard Schomburgk. Despite being aware that the local Indians considered their dead to be sacrosanct, he was determined to raid their final resting places in the name of science. With a partner in crime he dug up a skeleton of a Warao Indian, later presented to the Anatomy Museum in Berlin. On another occasion they obtained two recently buried Macusi Indian skeletons, both of which had been buried for no more than a year. During this particular instance of body-snatching the two were nearly caught red-handed and quickly had to hide both the skeletons and their digging tools under a bush until they could be safely retrieved later. Unlike his Russian counterpart in New Guinea, Schomburgk, as Rivière points out, seems to have had mixed feelings about what he had done and wrote that he 'was glad when the wicked work was finally and successfully accomplished'. In this act we can see a clear parallel with Burke and Hare, with the distinction that these mortal remains were snatched not to be dissected in a morgue but rather to be displayed in a museum.

Urry describes what is probably the most savage and appalling example of the immoral actions routinely pursued by civilised medical and scientific institutions on the mortal remains of natives. William Lanney and Truganini, described as the last of the Tasmanian Aborigines, had asked to be buried in peace when their time came. In 1869 Lanney died, and despite his wishes his corpse became the property of scientists. Urry gives an account of what took place next:

> In the morgue the body was viciously mutilated: the head, hands and feet removed and only the torso and limbs were left to bury. However, on the same night as the interment, two groups planned to exhume even these remains. Discovering their rivals had beaten them to the body, the leader of the other group smashed down the door to the morgue where the remains had been removed, only to discover 'a few particles of flesh' remained.

This pack of wild scientific dogs each carried off a piece of the corpse; one took an ear, another the nose, yet another a part of an arm, and the greatest prize of all – the head – was never seen again. A particularly sinister postscript to the story consists of the making of a tobacco pouch out of

Lanney's skin by Dr Stockwell, who was the Chief House Surgeon of the Colonial Hospital as well as a distinguished member of the Royal Society of Tasmania. It is impossible not to make parallels between this particular act and the hawking of combs made from Apache bones in the Wild West, as well as more recent examples, such as the making of lampshades from the skin of Jews under the Nazi regime. Truganini died in 1876 and was duly buried. But her skeleton was later dug up so that it could be displayed in the museum of the Royal Society of Tasmania. One can imagine Dr Stockwell admiring her skeleton whilst puffing away on his pipe with tobacco drawn from the pouch made from her countryman's skin. Finally in 1976 descendants of the Aboriginal Tasmanians regained control over her remains, which were then cremated and her ashes given up to the sea.

It was by no means the case that instances of body-snatching were restricted to the minor players in the development of anthropology. In fact many of the most important figures of this academic discipline on both sides of the Atlantic (whose work is still both admired in the profession and obligatory reading for students of anthropology) were involved. Even Franz Boas, one of the founding fathers of American anthropology, was implicated in such dubious goings-on. In the 1890s during the early part of his career he actually plundered native American graves and sold their skulls to subsidise his fieldwork. The morality of the actions of anthropologists of the British school was equally questionable. The leading Cambridge anthropologist Alfred Haddon even practised grave-robbing closer to home when he plundered a disused church in the west of Ireland, making off with a sack full of skulls. Such practices did not die out in the nineteenth century and possession by the spirit of science has led more recent scholars to commit acts of sacrilege. According to his own account, the palaeoanthropologist Don Johanson, best known as the man who found the remains of the australopithicene individual known as 'Lucy', together with a colleague named Tom Gray stole a femur from a family burial place in Ethiopia belonging to the Afar people. The reason behind this was that Johanson wanted a modern bone to compare with a fossil bone he had recently found, and for him, this seems to have provided sufficient justification for committing the act.

In this dark and morally dubious undercurrent of anthropology – having more in common with necromancy than science as normally understood – we can see in a microcosm the parameters of the whole strategy by which civilisation justifies its own savagery. The anthropologist, as an emissary of civilised man, can undertake such acts of grave-robbing in the name of science when he is dealing with the mortal remains of 'savages'. Were they attempting to do the same in the cemeteries of their own civilised cities they would be shunned as individuals devoid of moral conscience. If our moral superiority to 'primitives' and prehistoric 'cavemen' is to be cast in doubt by the bringing to light of such nefarious goings-on among scientists, what of the notion of social evolution and progress on a more general level? Can that be easily demonstrated? The belief in social and cultural progress is an essential component of the modern myth which would have us believe in a clear-cut distinction between prehistoric inertia and civilised momentum. Without underpinning civilisation with the notion of progress there would simply be no explanation for our claims for superiority over those outside the mainstream of human development (tribal societies) or before it (prehistoric societies). The belief in social progress is so integral a part of modern culture that it is often seen as an almost natural idea and as such has often been accepted unquestionably as a fact. In reality, it is largely an innovation of eighteenth-century Europe; before that it played a relatively minor role in human affairs. For many scientists and laymen alike the idea has an almost religious aura and attachment to it is often as fervent as the faith of religious believers.

The American anthropologist (not, to my knowledge, a grave-robber) Alfred Kroeber, rejected the notion that the sheer mass of knowledge that has accumulated throughout history could be considered as real evidence of progress. Much that passes for proof of social evolution can be dismissed as ethnocentricity. Yet even Kroeber could not let go of his own belief in the idea of progress. Whilst rightly dismissing the quantitative argument (i.e. that the inevitable accumulation of knowledge through history could be seen as evidence of progress in itself), he nevertheless sought to demonstrate a qualitative difference between modern civilised societies and other cultures. In order to advance this argument he argued that tribal cultures (and, by extension, prehistoric cultures) not only have no equivalent to our modern science, mechanics and technology but also have not 'evolved' beyond a reliance on magic and superstition – practices which we are said to have outgrown. In other words, we have something positive that they do not possess (science and technology) and they have something negative that we no longer possess (magic and superstition).

Leaving his point about science and technology to one side for a moment, let us look a little more closely at the question of magic and allied practices. Contrary to what many adherents of social progress would like to believe, magic, occultism and a whole host of what Kroeber would call superstitious practices are rife throughout the civilised world. As a belief system astrology may have been cast aside by the inexorable surge of modern rationalism and science, but if this is so, a great many people do not seem to have noticed it. Astrology has an enormous following, from the highest echelons of power (e.g. Nancy Reagan and members of the British Royal Family) to the readers of the hugely popular tabloid horoscopes. Avoiding walking underneath ladders, touching wood and the use of personal and team mascots are all practices as superstitious as anything found in tribal societies. As for magic itself, the old view of anthropologists on the subject, exemplified by Sir James Frazer's description of it as a 'bastard science', is now universally rejected. Many more recent anthropologists have seen magic and divination as coherent systems of belief. Sir Edward Evans-

Pritchard, who was Professor of Anthropology at the University of Oxford for many years, stated that the oracular form of decision-making used by the Azande people of Africa was 'as good a way of conducting one's affairs as any other I know'. The idea that magical thought is adhered to only by the most superstitious and poorly educated of people is, like the idea of social progress itself, a recent innovation. We need only look back as far as the seventeenth century to find the leading intellectuals of the day espousing magical and hermetic philosophies. The advance of modern science was achieved only at the considerable cost of the loss of much important magical research. As the philosopher of science Paul Feyerabend said: 'it is interesting to see that the demands of the new experimental philosophy that appeared in the seventeenth century eliminated not just hypotheses, or methods, *but the very effects* whose spuriousness was afterwards said to have been proved by scientific research'.

In that part of Kroeber's argument concerning the absence of anything comparable to the scientific and technological accomplishments of the civilised world, he is able to muster little to support his case that is genuinely of a qualitative nature. As we will see later in this book (Chapters 6 and 7), the emergence of scientific observations and modes of thinking can be traced back to remote times, and the development of simple mechanics and various technological procedures was not absent, even in the Old Stone Age. Somewhat ruefully, it appears, Kroeber was forced to conclude that science and technology were largely:

quantitative – like growth in population, growth in size of states and nations, growth sometimes in the number and complexity of their political subdivisions. Similar perhaps also is the growth in stocks of wealth – more gold, diamonds, farmed acres, houses in the world: part at least of the old are physically preserved, and new ones added. But we have already seen that there is serious doubt whether magnification as such, mere quantitative swelling-up, can be legitimately construed as progress. Size is easy to boast about, but does it bring with it wiser living or greater happiness? Only so far as it does can quantitative increase of culture be considered as making for progress.

Kroeber's argument that social progress has been achieved fails on both counts – due to the existence of scientific thought in the Stone Age and to that of magical thought in our own. Kroeber had a third point to add in support of progress which was essentially a corollary to his argument that the widespread belief in the efficacy of magic and divination is a mark of backward societies. This distinguishing feature of 'primitive' cultures concerned their inordinate interest in physiological and anatomical matters, which included the practices of deliberate head deformation, the filing of the teeth and the deliberate mutilation of the body for symbolic or cosmetic ends. Kroeber saw such practices as without parallel in civilised societies.

Circumcision and ear-piercing – both widespread practices in civilisations even today – were exempt from being included in his catalogue of primitive mutilations on the grounds that they were, 'after all, anatomically negligible'. Cosmetics, corsets and the like were similarly exempt, this time on the grounds that they did not cause any permanent defacement of the body or any mutilation. Were Kroeber alive today he would be very hard pressed indeed to divide primitive and civilised body practices so distinctly. How could he explain away even the currently quite common practice of body-piercing? How could silicon breast implants be described as impermanent alterations of the body – particularly as such operations rely on science and technology to make them possible?

Although much of what is called progress is simply the result of the accumulation of knowledge, this does not mean that there has been no loss along the way. The story of civilisation is one of both loss and gain. In rejecting the notion of a steady and uninterrupted progress in human affairs one is not obliged to advocate that human history is simply a chronicle of decline, a fall from a prehistoric Eden. Human achievements and failures through time occur in a discontinuous, irregular and altogether too complex and chaotic way to be reduced to either a simple form of progress or an equally simplistic doomsday theory of continuous and inevitable decline. Yet the supposedly linear nature of progress is often demonstrated by sleight-of-hand strategies.

For example, modern democracy is seen as a form of progress, one that has undoubtedly improved the rights and freedoms of the average citizen in comparison with, say, the earlier state of affairs that existed under the feudal system in Europe. Since feudalism preceded modern democracy in historical terms, then – so runs the argument – such progress is made manifest in the greater freedom of the individual. Often implicit in such arguments is the notion that since the serf had less freedom than the citizen of a modern civilised state, any individual in a society before feudalism would have had even fewer rights. That the average Stone Age individual may have enjoyed greater freedom than the serf (or even the average citizen of a modern democratic state) is simply ignored in this version of the human story, in which we ascend to ever greater heights and only look back in order to congratulate ourselves on how far we have come.

The notion of a straightforward and unequivocal progress also breaks down when we examine perhaps the most basic sign of any society's success – the state of health of its members. It is widely accepted that the hunter-gatherer people of the Old Stone Age did not, on average, live as long as we do today. But this fact partly depends on who 'we' refers to. Mark N. Cohen, in his book *Health and the Rise of Civilisation*, notes that contemporary hunter-gatherers (many of whom live in marginal environments, such as the Eskimos in the Arctic and the Kalahari !Kung bushmen in the desert) enjoy higher levels of calorific intake than the citizens of many Third World countries, and that these levels are also above those of the urban poor in 'advanced' societies. Cohen also makes the point that we have built up our images of human history too exclusively from the experiences of privileged classes and populations. By doing so we have assumed too close a fit between technological advances and progress for individual lives. A simplistic model of progress would have to demonstrate that there was a general and marked improvement in health as mankind left the hunting life for that of the farmer, and again when the industrial era began. In order to put the first of these assumptions to the test, Cohen and his colleague George J. Armelagos organised a special conference on palaeopathology (the study of disease in ancient and prehistoric human remains). Gathering together specialists in the palaeopathology of diverse regions of the world, they wanted to find out whether, with the shift from hunting to farming, the populations in the respective areas increased their general level of health or not.

The study of the sample of skeletal remains from South Asia showed that there was a decline in body stature, body size and life expectancy with the adoption of farming. A broadly similar result was obtained by the analysis conducted on skeletons from prehistoric populations in Georgia, USA – i.e. the health of the hunters was markedly better. In the case study of the Levant region there was a slight increase in the level of health with the initial adoption of farming, but this was followed by a marked decline once intensive agriculture and husbandry were fully established. Of the 13 regional studies, 10 showed that the average life expectancy *declined* with the adoption of farming. There are a number of factors that would have led to this decline. The domestication of animals that took place on a major scale with the advent of farming had, along with its benefits, the unforeseen result of allowing the transmission of numerous infectious diseases from these domesticates to their human masters. Among the side effects of the new lifestyle was the development of a new host of diseases and disorders, including beri-beri, rickets, leprosy (thought to have been transmitted to man from the Asian water-buffalo) and diphtheria. Diphtheria is one of at least 30 distinct diseases which can be transmitted via milk. As the palaeopathologist Don Brothwell has said, the practice of dairy farming undoubtedly assisted in the spreading of such diseases.

The next 'great leap forward' – industrialisation – also brought with it a whole mass of unwanted side effects, the so-called 'diseases of civilisation' precipitated by numerous features of the modern urban lifestyle, from sedentary habits and poor diet to environmental pollution. Obesity, heart disease, diabetes and cancer are among the disorders that are rare in earlier times and all too prevalent today. Palaeopathologists have conducted detailed examinations of hundreds of mummies from ancient Egypt, Peru and Alaska, and have found no traces of cancer. It also appears that the pressures of modern life have had a detrimental effect on mental health and may have precipitated numerous behavioural disorders as well. Thus the apparent advances made by modern civilisation often have their disadvantages. There are side effects in the most benign of breakthroughs, for even an advance in medical knowledge can be, and often is, diverted to military use.

In the light of the preceding pages it can be seen that the notion of progress is not as simple and straightforward as it may at first appear. It is still widely believed that the primitive and stagnant nature of Stone Age society was rudely awoken by the sudden appearance of civilisation 5,000 years ago. The entrenched belief that there was a quantum leap forward at this time has obscured the nature of prehistoric cultural activity by portraying it as simply consisting of a series of lowly steps on the ladder to true progress. The idea of biological evolution is distinct from the notion of social progress, although the two are often linked. This is inevitably the case with the study of our prehistoric origins. Accounts of prehistory necessarily have to explain how we evolved from our primate ancestors, and how we eventually became civilised human beings. This book does not seek to bring into question the idea of biological evolution. Its aim is to demonstrate, on the basis of archaeological evidence, that civilisation did not appear suddenly around 5,000 years ago. In order to do this, it is first necessary to give a brief standard account of the Stone Age – that vast time period from the origin of early man to the origin of the historical civilisations.

The Stone Age is so called because the main material used for making tools throughout this vast period of prehistory was stone. Archaeologists divide the Stone Age into three main periods – the Palaeolithic (Old Stone Age), the Mesolithic (Middle Stone Age) and the Neolithic (New Stone Age). The Palaeolithic period is by far the longest of the three, beginning with the earliest-known stone tools in Africa (2.4 million years ago) and ending about 10,000 years ago. As the time period involved is so long, prehistorians have found it useful to subdivide the Palaeolithic age into three periods – Lower (ending 200,000 years ago), Middle (200,000 to 40,000 years ago) and Upper Palaeolithic (40,000 to 10,000 years ago). The Lower Palaeolithic period was the time when our earliest ancestors (known as hominids) lived. All hominids belong to the family Hominidae and are divided into those of the genus *Australopithecus* and those of the genus *Homo*. Fossil remains of *Australopithecus* from East Africa date to at least three million years ago, and perhaps much earlier.

The first tools, however, are believed not to have been made by *Australopithecus* but by *Homo habilis* (who lived from about 2.2 to 1.6 million years ago), the earliest-known member of our own genus. Neither *Australopithecus* nor *Homo habilis* fossils have been found outside of Africa. The stone tools associated with *Homo habilis* have been found in East and South Africa and are called Oldowan, as such artefacts were found at the world-famous site of Olduvai Gorge in Tanzania. Fossil remains of this species are rare, and the presence of this species at archaeological sites is normally indicated by tools alone. The next species to develop in Africa was *Homo erectus* (1.6 to 0.5

million years ago), who possessed a larger brain than his predecessor. *Homo erectus* is generally accredited with being the first hominid species to leave the African homeland (about a million years ago) and colonise the more temperate zones of Asia and Europe. A more sophisticated stone tool technology known to archaeologists as the Acheulian (named after the discovery of such artefacts at Saint-Acheul in France) was developed by *Homo erectus*. The making of Acheulian tools was continued by the archaic populations of *Homo sapiens* who first emerged about 500,000 years ago.

The Middle Palaeolithic period saw further developments in both hominid evolution and tool-making. Neanderthal man (*Homo sapiens neanderthalensis*) arose about 100,000 years ago in Europe and western Asia as a regional development of archaic *Homo sapiens*. Contrary to the popular image of Neanderthals as the embodiment of brutish ignorance, they possessed large brains as well as being physically robust. The artefacts of this period are described as Mousterian (after the Le Moustier rock shelter in France) and are typically associated with Neanderthals. There is still considerable controversy concerning the ultimate fate of the Neanderthals, who disappeared off the face of the earth about 33,000 years ago. Some authorities argue that they were unable to compete with the incoming *Homo sapiens sapiens* and so became extinct, whilst others claim that they merged, at least in part, with our own immediate biological ancestors. Anatomically modern *Homo sapiens* first emerged in Africa at least 100,000 years ago and – like *Homo erectus* long before them – spread from their homeland to populate the world.

The Upper Palaeolithic period began 40,000 years ago, and with the decline of the Neanderthals, it saw *Homo sapiens sapiens* as the sole surviving member of the hominid line. Many experts see the onset of the Upper Palaeolithic as the time when behaviourally modern humans emerged. For although no major biological transformation of our own species took place 40,000 years ago, many see it as a time of a great leap forward in cultural terms, with an explosion of creative energies on all fronts. This cultural 'big bang' is seen as having included the birth of art, magic and religion as well as causing rapid advances in technology and social organisation. As in the earlier phases of the Old Stone Age, the vast majority of the artefacts from the Upper Palaeolithic are stone tools. Archaeologists have been able to distinguish distinct industrial traditions in the various phases of the Upper Palaeolithic. In Europe the main subdivisions include the Aurignacian, Gravettian, Solutrean and Magdalenian periods. Traditionally most prehistorians have been happy to see the Upper Palaeolithic period in terms of cultural progress, with the later phases showing higher degrees of technological, social and artistic development than the earlier ones. The Upper Palaeolithic ended with the last Ice Age, about 10,500 years ago, and was followed by the Mesolithic period. The Mesolithic is often described as a transitional phase between the old Palaeolithic lifestyle of hunting animals and gathering plants and the period of Neolithic farming. Changes that took place during the Mesolithic age include an emphasis on fishing and the development of new tools, particularly for use in woodworking. In outlying areas such as Britain, the Mesolithic period continued until the fourth millennium BC when the farming life characteristics of the Neolithic finally took root.

The Neolithic period began about 10,000 years ago in the Near East and is characterised by food production. In the Neolithic, the mainstay of the economy was neither hunting nor gathering but farming. In the main, the hunting of wild animals was replaced by the keeping of domestic livestock and the gathering of plants was supplanted by the cultivation of crops. The new Neolithic economy necessarily involved a settled way of life. The often highly mobile societies of earlier times were replaced by communities living in villages and towns. The making of pottery is often seen as one of the key diagnostic traits of the Neolithic, and many of the cultures of this period are named after their distinctive forms of pottery. The advent of farming was once described as the Neolithic Revolution, although archaeologists now realise that this is not accurate. The term 'revolution' suggests a dramatic and sudden event, but the change from hunting to farming took place over thousands of years. Towards the end of the Neolithic period, copper metallurgy emerged in some regions, and this has led some archaeologists to distinguish a Copper Age (or Chalcolithic period) from the preceding Neolithic and subsequent Bronze Age (the latter of which lasted in Europe from 2000 to 700 BC). Although the terms Palaeolithic, Mesolithic and Neolithic are used to describe the sequence of cultural events in Europe and parts of Asia, these terms are not generally used in discussing the prehistoric archaeology of other parts of the world such as the Americas and Australia. Although the evidence clearly points to the fact that both Australia and America were populated from Asia during the Upper Palaeolithic period, what happened subsequently on those continents cannot be adequately explained within the terminology of Old World archaeology.

From this foregoing account we can trace the conventional view of human progress from its humble origins to the advent of civilisation. According to the standard view, the hominids of the Lower and Middle Palaeolithic periods exhibit tool-making abilities far more developed than those of other primates but their expression is almost wholly limited to the utilitarian sphere – they are perceived as lacking the capacity for symbolic thought, as devoid of artistic ability, and without religious sensibilities. It is only with the cultural 'big bang' 40,000 years or so ago that behaviourally modern humans emerge as fully human and with artistic and religious awareness. The Neolithic period is the next significant stage of development, in which food production, pottery and other technological developments and urban settlements are added to the human repertoire. Yet even the most developed Neolithic communities are seen as lacking the essential ingredient of civilisation, namely writing.

The problem with this standard view of the Stone Age is that it does not adequately explain how the historical civilisations emerged out of this 'primitive' prehistoric heritage. If mankind before the historical era was so primitive, how could civilisation have arisen from such poor cultural roots? Historians of the ancient civilisations have, on the whole, paid little interest to the prehistoric background of the cultures they study, and as a consequence many wild theories claiming to explain the origins of civilisation have arisen. Archaeological theorist John R. Cole has described these highly unorthodox alternative explanations as 'cult archaeology'. The most famous of such theories in modern times is undoubtedly that of Erich von Däniken's *Chariots of the Gods?* (1970), in which the author claims that the sudden emergence of civilisation was due to alien intervention. Other writers who, like von Däniken and his numerous imitators, find it impossible to believe that 'primitive' Stone Age cultures could ever have developed into civilisations have sought the answer in theories concerning sunken continents. The myth of Atlantis is, of course, the most popular of these, but there are many other lands in the occult-pseudo geography of these theories, such as Lemuria and Mu. In both the extraterrestrial and the lost-continent models the basic theory is the same. Lost or hidden civilisations – be they alien or Atlantean – existed before those of Egypt and Mesopotamia and taught the latter everything they knew. The 'evidence' for such prehistoric civilisations is provided not by the physical remains of aliens nor by the archaeological remains of Atlanteans, as neither, of course, exist. Rather it is argued that the astronomical knowledge and the advanced technology of the ancient world *obviously could not* have been inherited from Stone Age cultures and *therefore* can only be explained by recourse to Atlantis or to aliens. Such views are extremely popular and influential, and this is partly due to public dissatisfaction with the standard academic view that does not explain the origins of civilisation in a convincing way.

I will show that the cultural elements that constitute civilisation did exist in the Stone Age and that the civilisations of ancient Egypt and other ancient societies had their prehistoric precedents. The evidence for the existence of civilisation in the Stone Age is given in this book. I take as my starting point the origin of civilisation in ancient Egypt, and from there go progressively further back in time to explore the body of evidence that clearly shows that all the elements of civilisation – writing, scientific thought and practice, medical knowledge, technology and art – were present in the Stone Age. Each of these elements is treated individually in turn and this necessarily involves shifting forwards and backwards in time. Nevertheless, the overall trajectory of the text is from the end of prehistory to its beginning.

Chapter 1

Stone Age Civilisations

Plate I shows one side of the well-known Palette of Narmer, from Hierakonpolis, Upper Egypt, now in the Cairo Museum. This singular artefact is a stock inclusion in books on the origins of civilisation, for it depicts the unification of Upper and Lower Egypt and heralds the beginning of the Dynastic period of ancient Egypt. It is literally an icon of the civilising force and has been described as the first page of the first chapter of the written history of Egypt. The large figure depicted in the centre is King Narmer (usually identified with Menes, the first king of the First Dynasty); in his raised left hand he holds a mace, emblem of his power and authority. He is in the act of subjugating his kneeling victim with a blow to the head. The two figures at the base of the composition represent previously dispatched victims. And the rest is history.

With the advent of history in Egypt there was clearly no great leap forward in terms of civilised behaviour, and the conflicts that existed in the Predynastic period simply reached a new level. How did this civilisation emerge – did it grow mysteriously out of an indigenous state of savagery, or was there some Atlantis-like civilisation which preceded it that professional archaeologists have overlooked? Such questions could be asked concerning any of the supposedly sudden arrivals of civilisations on the world stage, but of all the civilisations of the ancient world it is that of Egypt which draws the popular imagination like a magnet. The Pyramids of Giza and the enigmatic Sphinx are the launching pad for a thousand speculations into the secrets of lost civilisations. A passage in a recent work entitled *Keeper of Genesis: A Quest for the Hidden Legacy of Mankind* by Robert Bauval and Graham Hancock highlights the clash between cult archaeology and the standard academic view:

When was the 'genesis' of civilisation in Egypt? When did 'history' begin?

According to T.G.H. James, formerly Keeper of Egyptian Antiquities at the British Museum, and a representative voice of orthodox opinion on these matters: 'The first truly historical period is that which begins with the invention of writing and it is generally known as the Dynastic Period [...] the unlettered cultures which flourished in Egypt before the beginning of the Dynastic Period, and which exhibit some of the characteristics which mark the earliest phases of Egyptian culture in the Dynastic Period, are known as Predynastic ... Such traces of human life as are found in the Nile Valley dating from before the Predynastic Period are usually described in the terms used for European Prehistory – Palaeolithic, Mesolithic and Neolithic.'

So there we have it. Egyptian history – and civilisation with it – began at around 3100 BC. Before that there were merely 'unlettered cultures' (admittedly with some 'civilised' characteristics), which were in turn preceded by 'Stone Age' savages ...

The way James puts it, the whole picture seems very clear-cut, orderly and precise. He really makes it sound as though all the facts are now in hand concerning the Predynastic Egyptians and their forebears, and that nothing more remains to be discovered about any of them.

Such anodyne notions of the past are widespread amongst Egyptologists who again and again in their textbooks ... convey the comforting impression that the prehistory of Egypt is well understood, organised, categorised and safely put in its place ...

But is everything really as orderly and as well worked out as the 'experts' say? And can we really be so sure that there is 'no big surprise in store for us'?

The tenor of Bauval and Hancock's view is clearly expressed in this passage: orthodox Egyptologists do not supply us with a convincing explanation for the genesis of civilisation in Egypt. These authors then provide us with an alternative explanation which belongs to another world of thought from that represented by the academic tradition, one that is more a modern reworking of the Atlantis myth than a reappraisal of the Predynastic era in Egypt. We need not follow either the historians of the ancient world who seem to feel that there is no real need to explain the apparently abrupt rise of civilisation in Egypt, or take recourse to the equally unsatisfactory speculative approach which wishes to build castles in the sands of Egypt.

Traditionally Egyptologists who sought to explain the sudden transformation to the civilised state marked by the beginning of the Dynastic period believed that it could be best explained by a rather dubious notion of a 'Dynastic race'. This incoming élite were supposed to have swept into Egypt and to have transformed the prehistoric barbarism that greeted them on their arrival. This idea has all but died out, but even in the mid-1950s D.E. Derry was still propagating such a belief and the racist assumptions underlying it. He compared measurements he had taken from Predynastic and Dynastic skulls and decided that the Predynastic inhabitants of Egypt were probably of 'Negro stock' whilst the Dynastic skulls indicated that the incoming (white) race had 'a greater cranial capacity and of course a larger brain'. William Arnett, who has investigated the history of this totally fallacious theory, quotes the conclusions reached by Derry based on his examination of the Dynastic skulls:

> It is ... suggestive of the presence of a dominant race, perhaps relatively few in number but greatly exceeding the original inhabitants in intelligence ... a race which brought into Egypt knowledge of building in stone, of sculpture, painting, reliefs, and above all writing; hence the enormous jump from the primitive Predynastic Egyptian to the advanced civilisation of the Old Empire.

There are few Egyptologists who would share such views of an exotic master race storming into Egypt and setting up their civilisation as if it were a prefabricated building. Whilst there is little doubt that there were significant external influences on Egypt from the Near East at this time, a growing number of researchers believe that the main developments grew out of indigenous Predynastic traditions. Contrary to what Bauval and Hancock suggest, specialists in the study of the Predynastic period do not think that 'all the facts are now in hand' concerning this era. It is becoming increasingly obvious that there was no 'enormous jump' as Derry and many others thought. Rather than being an explosive act of foreign intervention, the archaeological evidence indicates that the emergence of Egyptian civilisation is better seen as the gradual outgrowth of the indigenous prehistoric culture. The final phase of the Predynastic, known as the Gerzean, began about 4000 BC and pre-empted many of the features of the Dynastic period. Gerzean tombs have been discovered that consist of several furnished rooms and quite clearly foreshadow the treatment of the dead in pharaonic times. Many religious motifs, including a number of themes associated with some of the gods of Dynastic times, also have been found to have prehistoric origins, and the distinction between civilisation and barbarism in Egypt is a very fine line indeed. This transition can no longer be seen as abrupt and this is particularly clear in the case of hieroglyphic writing.

Writing is, of course, one of the main features of those societies considered to be civilised, but in the case of ancient Egypt this too is now perceived to have its Predynastic precursors. It was once thought that the Egyptian system of writing drew its inspiration from Mesopotamia, particularly the Sumerian civilisation. This was always something of a problematic idea because hieroglyphs bore little resemblance to Mesopotamian writing, but the alternative – that they had developed out of the barbarism of the Predynastic period – was considered altogether unlikely. A number of scholars have independently reached the conclusion that the hieroglyphs did indeed develop out of Predynastic roots. Design motifs on pottery and other artefacts from 4000 BC onwards have been shown to be precursors of hieroglyphs a thousand years later; even a number of basic letters of the hieroglyphic system have their origin in Predynastic figures.

The numerical system of the Egyptians was also highly developed before the Dynastic period. Narmer's mace-head displays the hieroglyphic symbols for the number 1,422,000, showing beyond any shadow of a doubt that major intellectual developments occurred before the supposed advent of civilisation. William Arnett, whose own study of the Predynastic origin of hieroglyphs has shown quite clearly that the Egyptians did not rely on external influences to develop their system of writing, has suggested that: 'If enough Predynastic inscribed material can be collected and catalogued, it may be possible to write an earlier chapter to the *history* of Egypt and to cast light upon an epoch which remains obscured in what, at present, must be termed *prehistory*.' Thus it cannot be ruled out that we may yet see the origin of hieroglyphic writing, and therefore Egyptian civilisation itself, being placed further back in time.

As Egyptian civilisation was once thought to have developed as the result of primarily Mesopotamian influence, so too Europe has been envisioned as having developed more as a result of Near Eastern influence than by drawing on its own prehistoric heritage. Civilisation in Europe has usually been seen as a rather late development, but recently there have been dissenting voices to this kind of thinking. The late Marija Gimbutas became an increasingly controversial and marginal figure in archaeology as her rewriting of European prehistory developed over the years. According to her account, between 7000 and 3500 BC the early farming communities of Europe coexisted with each other and with nature in a largely peaceful fashion, worshipping a Great Goddess. This period of stability and civilisation began to come under threat from the repeated invasions of Proto-Indo-European peoples from the east. During the period of these assaults, between 4300 and 2800 BC, the ancient cultural values of Old Europe gradually collapsed, only surviving in outlying areas and eventually disappearing underground, lasting into historical times only in a partial form in folklore and mythology.

In her work the old motto of V. Gordon Childe and many other archaeologists, *Ex oriente lux*, is brought into question, and Old Europe is shown to be culturally distinct and largely independent of oriental influences until the destruction of traditional values by the incursions from the east. Old Europe is seen to represent a civilisation in its own right. She argued that as Sir Arthur Evans had shown that Greek civilisation could not be understood without reference to the Minoan and Mycenaean cultures that preceded it, so these latter civilisations could only be fully assessed with a deep knowledge of the even earlier civilisation of Old Europe. She was dismissive of the need to account for the 'sudden' appearance of Minoan and Mycenaean civilisations by recourse to the legend of the lost civilisation of Atlantis. The spurious notion that the foundations of civilisation disappeared from view when Atlantis sank beneath the waves is replaced by a far more convincing argument. She shows that Old Europe was the precursor of many later cultural developments and that the ancestral civilisation, rather than being lost beneath the waves through some cataclysmic geological event, was lost beneath the waves of invading tribes from the east.

In the preface to her last major work, *The Civilisation of the Goddess*, Gimbutas boldly wrote:

The use of the word *civilisation* needs an explanation. Archeologists and historians have assumed that civilisation implies a hierarchical political and religious organisation, warfare, a class stratification, and a complex division of labor. This pattern is indeed typical of androcratic (male-dominated) societies such as Indo-European but does not apply to the gynocentric (mother/woman-centred) cultures described in this book. The civilisation that flourished in Old Europe between 6500 and 3500 BC and in Crete until 1450 BC enjoyed a long period of uninterrupted peaceful living which produced artistic expressions of graceful beauty and refinement, demonstrating a higher quality of life than many androcratic, classed societies.

I reject the assumption that civilisation refers only to androcratic warrior societies. The generative basis of any civilisation lies in its degree of artistic creation, aesthetic achievements, nonmaterial values, and freedom which make life meaningful and enjoyable for all its citizens, as well as a balance of powers between the sexes. Neolithic Europe was . . . a true civilisation in the best meaning of the word. In the 5th and early 4th millennia BC, just before its demise in east-central Europe, Old Europeans had towns with a considerable concentration of population, temples several stories high, a sacred script, spacious houses of four or five rooms, professional ceramicists, weavers, copper and gold metallurgists, and other artisans producing a range of sophisticated goods. A flourishing network of trade routes existed that circulated items such as obsidian, shells, marble, copper, and salt over hundreds of kilometres.

In this passage Gimbutas alludes to a number of aspects of the cultural life of Europe during the New Stone Age in which she sees evidence of a different kind of civilisation. According to her, this civilisation was based on a social order in which men and women had equal status. The religious life of Old Europe centred on the worship of a goddess who took many forms. The earth was revered as the embodiment of the goddess and death seen as a return to the womb of the earth/goddess. Tens of thousands of figurines and other representations of the goddess, along with highly ornate pottery, have been unearthed from the archaeological sites of Old Europe, thus giving us a very detailed picture of the artistic and religious traditions of these communities. The holistic ideology of Old Europe was challenged and eventually overrun by the forefathers of the Indo-Europeans, the late Neolithic Kurgan culture that originated on the steppes of Russia. The Kurgan culture, with its horse-riding warriors armed with lethal weapons, brought into Europe an ideology based on patriarchy, hierarchy and military prowess. Their pantheon of gods was male-dominated and headed by a sky god. The Earth Goddess and other female deities were, under the new order, reduced in status to become merely the wives of the gods. Sexual inequality, militaristic violence, dualistic thinking and a fundamental belief in linear continuity (all features integral to our own civilisation) are found in Kurgan culture and amongst the later Indo-Europeans. The Stone Age philosophy of Old Europe, with its emphasis on cyclic time and holistic social and ecological thinking, was pushed aside as the new ideology gained ascendancy. The beliefs of Old Europe survived as an undercurrent but the foundations of a new and savage civilisation were beginning to be built.

Gimbutas' reconstruction of the cultural ideas and practices of Stone Age Europe has gained her a reputation that extends far beyond the community of academic archaeology. Feminists and New Age thinkers have seized upon her work as a fecund source for blueprints for a new vision of society. Unfortunately many such supporters have lacked her critical acumen and misrepresented her ideas, and this has hardly helped her reputation amongst archaeologists. Yet even without her having drawn support from such (to academics) dubious quarters, her radical vision of Old Europe and its inbuilt critique of what we generally assume to be civilisation would inevitably have led to contention with her professional colleagues. For despite her impeccable academic work and immense erudition (no other individual scholar has made such an in-depth study of the artistic traditions of eastern European prehistoric cultures), her ideas do not sit well with current views concerning civilisation. Many – mainly male – prehistorians have violently attacked her ideas as being based on a somewhat romantic view of the past, and deny that peace or sexual equality were ever widespread social features in the cultural landscape of Stone Age Europe.

In fact, her identification of the Kurgan culture as Proto-Indo-European *is* open to debate, as the whole question of the homeland and subsequent migrations of the Indo-Europeans is yet to be settled. Whether the ancestors of the Indo-Europeans can be blamed for destroying the earlier civilisations of Neolithic Europe is therefore something of a moot point. Similarly, it can be said that Gimbutas saw the iconography of the goddess everywhere and perhaps inferred too much about the social order of Old Europe on the basis of too little evidence. Yet even if one remains unconvinced by some aspects of Gimbutas' reconstruction of events in prehistory, it is apparent that in Stone Age Europe and beyond there were many complex and sophisticated cultures. That these cultures are best described as civilisations is a view shared by other archaeologists, as will be made clear in the next few pages. Many of her insights into the symbolism of Neolithic religions are firmly supported by the evidence and she cannot be dismissed simply as a maverick. This period of prehistory was one of great dynamism and complexity and one that was by no means purely dependent on external influences for its innovative spirit and great artistry. Many of the technological, artistic, symbolic and even scientific and medical aspects of Neolithic and earlier European and other cultures are dealt with at length in subsequent chapters of the present book, and so for the moment I shall limit myself to a brief account of temples and other examples of sacred architecture from prehistoric archaeological sites that attest to the rich spiritual life of our Stone Age forebears.

The site of Çatal Hüyük, 32 miles south-east of Konya in the southern part of central Anatolia (present-day Turkey), is one of the most spectacular examples of Neolithic civilisation yet discovered. Covering an area of more than 20 hectares, this large Neolithic town – some, including its excavator James Mellaart, have described it as a city – has been estimated to have supported a population of 7,000 people. Founded more than 8,000 years ago Çatal Hüyük seems to have been a thriving community for a thousand years or more. It is the largest human settlement site yet discovered from the Neolithic era. It is a remarkable fact that the largest town known from the Stone Age should have belonged to the earlier part of the Neolithic period rather than towards its end when, if one adheres to a simple model of progress, it would be expected to have arisen. The striking nature of the site prompted Mellaart to exclaim that 'the Neolithic civilisation of Çatal Hüyük represents something unique in the long history of human endeavours: a link between the remote hunters of the Upper Palaeolithic and the new order of food-production that was the basis of all our civilisation', and again: 'Neolithic civilisation revealed at Çatal Hüyük shines like a supernova among the rather dim galaxy of contemporary peasant cultures.' Future discoveries may well reveal that other bright stars once shone in the firmament of Neolithic civilisation both in Anatolia and beyond.

Stone was in short supply in the environs of the settlement and the buildings were not made of stone at all but of sun-dried mud brick. The rectangular houses of Çatal Hüyük had timber frames and flat roofs with access via openings in the roof. Numerous shrines were found amongst the buildings, and plaster reliefs and wall paintings provide a fascinating glimpse into the social and religious life of the town's inhabitants. *Figure 1* is a plan of part of the settlement which shows how the shrines (marked with an S) were an integral part of the cultural life of the town. *Figure 2* shows a section of the terraced houses and shrines of the settlement, with ladders allowing easy access between the various levels. Among the numerous remarkable discoveries at the site is a Stone Age 'map' of the town, painted on the walls of one of the shrines. In the foreground of this map the characteristic terraced housing of the town is depicted. Behind the town is an image of an erupting twin-peaked volcano spewing lava, smoke and ash. This represents an actual volcano visible from Çatal Hüyük which, although now extinct, is known to have been active during the Neolithic period. This volcanic mountain was the main local source of obsidian (black volcanic glass), which was highly sought after for making tools on account of its extremely sharp cutting edges. Mellaart has suggested that in

Neolithic times the volcano was surrounded with a magical aura, for not only was obsidian a gift from mother earth but it was also intimately associated with fire and the underworld. Thus this painting of a prehistoric eruption is the recording not merely of a spectacular natural event but also of an event seen to have supernatural overtones.

Obsidian was of great economic importance in Neolithic times, and trading in this raw material seems to have been a major factor in the size and prominence of Çatal Hüyük. The mainstays of the agricultural economy of the town were the cultivation of wheat and barley and the domestication of cattle. The hunting of wild animals also played an important role in providing meat for the townspeople and had by no means been entirely supplanted by the practice of farming. The quality of the pottery, tools, personal adornments, fabrics and various other types of artefacts (including the world's earliest mirrors, made from polished obsidian) from the site shows the high levels of craftsmanship which are indicative of craft specialisation. One especially striking example of the industrial activities at Çatal Hüyük is a wall painting which is clearly an imitation of a textile pattern. It has a geometric pattern incorporating floral motifs and was painted in light buff, white, grey, black and orange red. This shows that woven rugs that we now know as *kilims* were innovations of the Stone Age and that the history of the oriental rug goes back to a period much, much earlier than the Islamic era.

The complex religious iconography of the site is revealed in the wall paintings, sculptures, statuettes and other remains from this early Neolithic town. Mellaart has shown that the religion and mythology of this civilisation was built around the worship of a Great Goddess. His interpretation, although worked out before Gimbutas formulated her own reconstruction of the Neolithic goddess religion, is in harmony with her findings. He believes that it is extremely likely that most of the rituals that took place at Çatal Hüyük were conducted by priestesses and that male priests played a comparatively minor role in the ceremonial life of the community. As the biological source of life, the Neolithic woman became a symbol of the agricultural enterprise – of fertility, seasonal renewal and both life and death. Those aspects of the goddess that were Neolithic in origin were grafted on to earlier branches of mythology associated with the hunting life that was universal in the Upper Palaeolithic period. These more archaic elements of the goddess mythology were celebrated at Çatal Hüyük by depictions of her in association with wild beasts, most strikingly with the most dangerous predator of the region – the leopard. A clay statuette found in a shrine at the site portrays the enthroned goddess resting her hands on two leopards; she is giving birth, perhaps, as Mellaart suggests, to a male child. Male images do appear in the iconography of the shrines but are usually, but not always, subsumed by the powerful figure of the goddess.

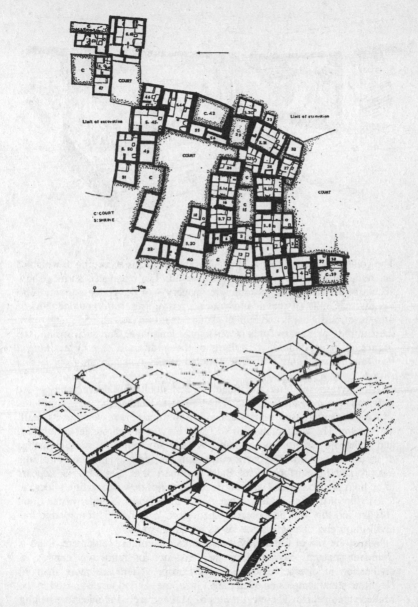

Figures 1 and 2

Figure 3

The goddess herself is portrayed in a number of forms according to which of her numerous aspects was to be emphasised. The excavated shrines show such a bewildering array of dramatic imagery – of the goddess giving birth to rams and bulls, of actual vulture beaks protruding from moulded pairs of breasts, of bull's horns and human skulls – that one can only be staggered at the complex grammar of this mythological language. One such shrine (see *Figure 3*) shows vultures (emblems of the goddess of death) attacking a headless human figure with disembodied skulls lying on the floor of the shrine.

The mysteries of life and death in this ancient goddess religion are mirrored in the history of Çatal Hüyük itself. Neither the earlier cultures which gave birth to this remarkable civilisation nor the reasons for its death are known. The decline and fall of Çatal Hüyük are shrouded in mystery, and archaeologists have been unable to trace any real discernible influence of its civilisation on subsequent cultures in the region. This state of affairs has prompted prehistorian Paul Bahn to remark that 'Çatal Hüyük appears to be both precocious and without issue.' A major new excavation under the leadership of the Cambridge archaeologist Ian Hodder is underway at Çatal Hüyük, and this will undoubtedly shed more light on this remarkable lost civilisation and untangle some of its ancestral roots.

Whilst the site of Çatal Hüyük is a unique Neolithic settlement, there is abundant material evidence for the existence of numerous centres of civilisation in diverse regions of Old Europe. There are more than 40 Neolithic stone temples on Malta and Gozo, the second largest island in the Maltese archipelago. These examples of Maltese sacred architecture belong to the period 4500 to 2500 BC and demonstrate how a people without the use of any metal tools were able to construct monumental buildings using blocks of stone weighing as much as 50 tons. At one of these sites, the 5,000-year-old temple complex at Tarxien, the surviving part of a monumental stone statue was discovered. Only the bottom part remains. It depicts the lower legs and part of the skirt of what seems to be a standing female figure. As this part is a metre high, the original height of the statue has been estimated at nearly three metres, making it the earliest known example of a monumental statue on this scale. *Plates II and III* show the temple of Hagar Qim (3500 to 3000 BC) on the southern coast of Malta. Sculptures and figurines found inside the temple clearly show the importance of the female figure; among them is a finely made terracotta figurine of a naked woman which has come to be called the 'Maltese Venus' (see *Plate IV*).

Of equal dramatic power to the major Neolithic temples of Malta is the Hypogeum of Hal Saflieni (built in stages from 3600 to 2500 BC), near the Tarxien temple. Hypogeum is the name given by archaeologists to rock-cut chamber tombs, and the complex at Hal Saflieni is a remarkable example of this kind of monument. Hal Saflieni is a conglomeration of numerous simple single tombs that have been linked by passages, stairways and subterranean halls. This labyrinthine complex covers 480 square metres and was cut into the top of a limestone hill using stone mallets and horn or antler picks. Once the rough shape and size of each chamber had been achieved by the use of these tools, the walls were smoothed with implements made of flint. Excavations have revealed three storeys of interconnecting chambers, and there are deeper levels that remain unexcavated. There are more than 30 chambers in the three storeys, most of them in the middle level. The lower two levels have been carved and decorated with complex spiral and other patterns painted in red. A number of chambers on the second storey have walls shaped in imitation of the megalithic architecture of the island (see *Plate VI*). As in many of the temples, female figurines have also been found in the Hypogeum. One such work of art, which has been hailed as a masterpiece, is a 12-centimetre-long terracotta figurine known as the 'Sleeping Lady' of the Hypogeum (*Plate V*). The serene pose of this dreaming woman may seem somewhat out of place in a complex of tombs, but its presence there has led Gimbutas to suggest that initiation rites involving symbolic rebirth may have been conducted in the Hypogeum.

The archaeologist J.D. Evans made a detailed study of the Neolithic cultures of Malta and expressed the opinion that:

insofar as we can judge from the evidence, no more peaceable society seems ever to have existed. It is easy, of course, to delude oneself with pictures of a primitive Mediterranean paradise; nevertheless, the earth seems to have yielded the primitive Maltese a living on fairly easy terms, for otherwise they would scarcely have had time or energy to spare to elaborate their strange cults and build and adorn their temples.

But the peaceful and bountiful life enjoyed by the temple-builders came to an abrupt end around 2500 BC. The reasons why this flourishing and vibrant civilisation disappeared are as obscure and mysterious as those that resulted in the end of Çatal Hüyük. Evans, noting that the 'temple-builders vanish as if by magic', sought to explain this by an invasion of war-like people. Gimbutas has suggested that perhaps the natural resources of the island could no longer sustain its inhabitants and deforestation or crop failure may have brought famine, disease and other disasters in its wake. Whatever the reasons for their disappearance, the magic of the temple-builders endures in the monuments they left to posterity.

The art from both the shrines of Çatal Hüyük and the temples of Malta attests to the central importance of female deities during the Neolithic period. Various regional traditions of Old European architecture and burial practice also display a preoccupation with the same themes. Symbols such as the uterus, the vagina, the pubic triangle, the egg and in some cases the whole female body were all used to express these spiritual beliefs. One of the most striking examples of the use of the symbol of the egg comes from the 7,000-year-old cemetery at Nitra in Slovakia. Each of the graves at the site is egg-shaped, and this may well have expressed the belief that the dead were to be reborn after having returned to the earth goddess. In other regions, such as the British Isles, the goddess seems to have been literally embodied in the architectural plans of various Neolithic constructions. The entrance to Norn's Tump, a 5,000-year-old long barrow in Gloucestershire, England, has been likened by Gimbutas to a vulva (*Figure 4*). Looking at this picture as a whole, the stone walls leading to the entrance may be seen as the open legs of the goddess, whilst the barrow beyond the vulva-like opening may represent her (pregnant) belly. A similar interpretation can be given for the structure and layout of numerous other Neolithic sites. The plan of the Neolithic tomb of Shanballyemond, Tipperary, Ireland (*Figure 5*) can be seen as a representation of the lower part of the goddess' body, with the central chamber as the uterus and the 'antechamber' as the vagina. A fourth-millennium BC megalithic tomb at Carrowkeel in County Sligo is distinctly anthropomorphic, or human-shaped (*Figure 6*), and as Gimbutas has noted seems to foreshadow the cruciform churches and cathedrals of the Christian era.

Among the remains at numerous Neolithic sites in eastern Europe, curious clay models of temples have been repeatedly unearthed. In fact hundreds of such artefacts have been discovered. They appear in the archaeological record from about 8,000 years ago and are one of the main sources of our information on the prehistoric religious architecture of the

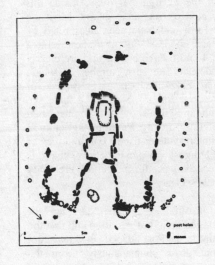

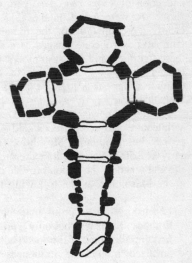

Figures 4, 5, 6

region. Their exact function is unclear, but Gimbutas suspects that they may have been votive offerings to a full-scale temple, or perhaps symbols of the goddess herself. The model shown in *Figure 7* is from Porodin in western Macedonia, belongs to the early Neolithic period and is about 8,000 years old. It is 35 centimetres in height and depicts the face of a goddess on its chimney stack and her necklace encircling the chimney on the roof. Another temple model from the same site was found to contain a miniature altar inside it. These models were probably replicas of full-scale temples that were made from perishable materials and for that reason have not survived in great numbers, unlike, for instance, the grand stone passage-graves and stone circles of Neolithic Britain and Ireland, or the temples of Malta.

Nevertheless, there are remains of dramatic shrines at sites in eastern Europe. A large prehistoric shrine (about 70 square metres) was discovered at the site of Sabatinivka II in Moldavia (western Ukraine) and has been assigned to the Cucuteni culture (*c.* 4800 to 4600 BC). The illustration (*Figure 8*) highlights a number of significant features of the shrine: (1) a flagstone floor at the entrance; (2) an oven; (3) a plastered platform or altar (2.75 × 6 metres); (4) a lifesize chair made from clay; (5) figurines; and (6) a collection of vases near the oven. Of the 32 figurines found inside this shrine, 16 were found on the altar seated on miniature chairs. Most are clearly representations of the female form and have been interpreted by Gimbutas as concrete images of a snake goddess. *Figure 9* shows six of these figurines, and whilst numbers 1, 4 and 6 have breasts and are clearly female, numbers 2 and 3 are of indeterminate sex. Number 5, which Gimbutas suggests is a female holding a 'snake-like infant', a baby snake or a phallus, could perhaps better be interpreted as a masturbating male.

The site of Lepenski Vir on the Danube in Serbia shows that sophisticated architecture and sculpture existed not only among Neolithic agricultural peoples but also among Mesolithic (hunter-gatherer-fisher) communities as early as 8,500 years ago. Its excavator Dragoslav Srejovic believes that this site and others in the region provide clear evidence for indigenous creativity in the Mesolithic, as the following makes clear:

Lepenski Vir is a proof that the Mesolithic is not the 'dark age' of European prehistory but only a prolonged period of gestation and that the first great advance of European culture in the Post-Glacial period was not fertilized by outside influences but arose spontaneously, from the awakening of the long-concealed energies of the Danubian culture of the Late Palaeolithic. This is a significant fact. It shows that Europe did not have to borrow from the Near East in order to rise above the past and find the strength for a creative future. Lepenski Vir is a proof that achievement took place in Europe independently. Its unexpected revelation and its exceptional features prove that our knowledge is still limited.

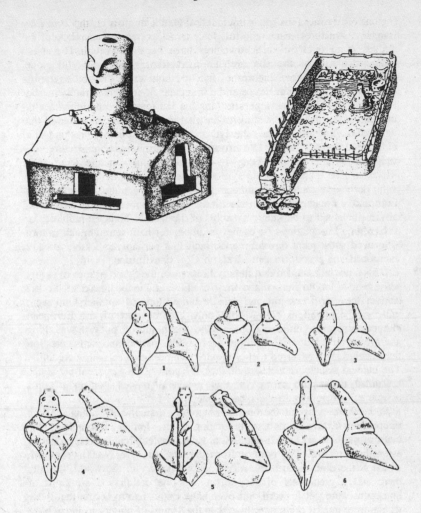

Figures 7, 8, 9

The site of Lepenski Vir consists of a considerable number of trapezoidal or triangular structures that are often referred to as sanctuaries, temples or shrines; more than 50 such structures have been excavated. The most striking features of the site are the extraordinary and powerful stone sculptures which merge features of fish and humans in their iconography (see *Plate VII*). Gimbutas has seen the triangular floor plans of the Lepenski Vir temples as deliberate representations of the female pubic region, with the two converging sides leading to the altar in the interior, an image of the uterus of the goddess. For her the symbolism of these temples was part of a widespread tradition across Old Europe in which religious architecture was designed to represent the body of the goddess in concrete form. The inspiration for these kinds of symbolic architectural traditions, far from being derived from Near Eastern archetypes, may be traced to indigenous roots and ultimately to the Palaeolithic decoration of caves, clefts and crevices believed to be sacred symbols of the womb. Thus the architecture of Neolithic Europe drew on earlier traditions for its inspiration and in turn influenced subsequent developments, including perhaps, as I have already mentioned, the cruciform cathedrals of Christian Europe.

It must not be assumed that the civilisations of the Stone Age were in any sense limited to Old Europe and the Near East, and indeed, across the other side of the world cultural and artistic innovations of another kind were underway, innovations that would only become part of the European cultural repertoire millennia later. The invention of pottery was long assumed to have belonged firmly in the Neolithic period, and any suggestion that it could have occurred earlier would have been treated with incredulity. The earliest widely accepted evidence for pottery in western Asia (once presumed to be the centre of this innovation) does not disturb this assignment of pottery to the New Stone Age.

In 1960 this view of the origins of pottery dramatically changed when radiocarbon dating of pottery belonging to the Jomon culture from the Natsushima site at Yokosuka, Kanagawa Prefecture, Japan, produced astonishing results. The pottery from the site was over 9,000 years old, which at the time made it the oldest known pottery in the world. Initially there was a great deal of scepticism over the dating and all kinds of objections were put forward, but eventually critics had to accept that these results were indeed accurate. Since then the origins of pottery in Japan have continued to be put further and further back in time. So far the earliest dates to around 12,700 before present (BP) but even this pottery, which is decorated, is clearly not the earliest attempt to produce ceramic vessels. The Stone Age Jomon culture that flourished in Japan until about 400 BC was not one that was based on a farming economy and it was thus something of a shock to archaeologists in both Japan and elsewhere to find such an advanced tradition of pottery in such seemingly 'primitive' cultural circumstances.

Even more surprising to some (even Japanese) scholars was the presence of the world's earliest pottery in Japan. Archaeologists had generally thought of Japan as dependent on China for cultural innovations, and even when the early dates of the Jomon pottery were shown to be authentic (and older than anything comparable from mainland Asia), many researchers assumed that it was just a matter of time before even earlier pots were found in China. This has not, at least as yet, occurred; the earliest pottery from north-east China is about 9,000 years old whilst that found in south China may be as old as 10,000 BP, although this latter date is still a matter of some controversy. Yet even if we accept this earlier date, it is still significantly later than the first pottery known from Japan. However, recent discoveries on mainland Asia may indicate another very early instance of the production of pottery – not from the traditions of China but from the so-called cultural backwater of eastern Siberia. Ongoing excavations at Stone Age sites in the Amur River region have revealed that pottery fragments found there may be 13,000 years old and thus as old as the earliest Japanese evidence. It is beginning to seem that pottery was an established tradition in Japan, China and Siberia at a very early date and further archaeological investigations are likely to reveal more about the extent and antiquity of the use of pottery in north-east Asia. Nevertheless, in the light of present knowledge, it is the so-called peripheral cultural zones of Japan and Siberia that have produced the earliest dates, and not China.

Whilst the Jomon culture had thus become rightly renowned for the great technological and artistic skill of its pottery and other ceramic artefacts, it was still seen as an inferior culture since it belonged firmly in the Stone Age. The Japanese themselves identified their ancestors as the people who belonged to the Yayoi period that came after the eventual demise of Jomon culture. A clear distinction was made between the uncouth and backward Jomon culture and that of the Yayoi people. The sequence of Japanese prehistory is very different to that of Europe. Thus the model of social change that was developed to explain the nature of European prehistory – namely the transition from Palaeolithic to Mesolithic to Neolithic to the Bronze Age and finally to the Iron Age – is entirely inappropriate for the case of Japan. The Jomon culture has features of both Mesolithic and Neolithic cultures and cannot be adequately described by either term. It is Mesolithic in that it was based on a hunting, gathering and fishing economy (rather than a farming economy, as in Neolithic cultures), but pottery is a diagnostic feature of a Neolithic lifestyle. Despite the Jomon being essentially pre-Neolithic and thus supposedly more 'backward', its extremely precocious ceramic tradition is of a technical and artistic quality rarely, if indeed ever, equalled in fully fledged Neolithic societies. The Yayoi period began about 400 BC and its distinguishing features were agriculture and the use of iron, both of which were imported cultural traits from China. Thus Japan shifted directly from the Stone Age Jomon culture to an iron-using culture, taking a radically different cultural trajectory from Europe. The agriculture practised by the Yayoi people was wet-rice or paddy-field farming, and the central role that rice plays in Japanese culture is undoubtedly a major factor for modern Japanese people to identify the Yayoi people as belonging to their race. The greater social stratification of the Yayoi period and the fact that the cultural foundations of the state were built at this time are also examples of the kind of cultural achievements of which a modern nation could be proud.

On the other hand, the Jomon culture apparently had no rice cultivation and only a simple social order and was seen as a primitive culture that had little but its pottery to commend it to the self-image of the Japanese. It was thus suggested that the Jomon culture belonged to an inferior race (this belief echoes the old and largely discredited theory that the inhabitants of Predynastic Egypt were of an inferior race to those of the Dynastic era). This denigration of Stone Age cultures and their contributions to later cultural developments is therefore hardly unique to Japan. The idea that Stone Age Europe was indebted to the Near East for almost all of its cultural innovations was a belief that distorted our understanding of European prehistory for far too long. As the prehistoric period in Europe has become better understood, many aspects of technology, artistic and religious heritage and culture as a whole have been seen to have had indigenous roots rather than simply to have been borrowed from more 'civilised' cultures further to the east. Now it seems Japan must reassess its prehistoric forebears in the light of dramatic recent discoveries concerning the Jomon culture. There are a number of archaeological finds that now suggest that rice cultivation may not have been an innovation of the Yayoi people at all but part of Jomon culture. Three-thousand-year-old shards (i.e. broken fragments) of pottery that have impressions of rice hulls on them were found at Minami-Mizote in Okayama Prefecture, and rice discovered at Kazahari (Aomori Prefecture) indicates that this was quite clearly not an isolated phenomenon. Even more telling has been the discovery of not only rice but also the vestiges of rice-paddies in an archaeological layer belonging to the Jomon era at a site in Itazuke, Fukuoka Prefecture. It seems then that rice may well have been cultivated before the Yayoi period, and by the Jomon people, who had previously been seen as too primitive to have achieved this, being as they were a Stone Age culture.

A far more dramatic reappraisal of Jomon culture has had to be made in the light of the recent discovery of the largest settlement known from this period. In 1992, during the course of surveying work undertaken for the planned construction of a new baseball stadium in Aomori City, it became apparent that there were significant archaeological remains dating from both the Jomon period and later times. As the scale of the Jomon settlement (known as the ruins of Sannai-maruyama) became apparent over the next two years, it was realised that the site was the most important archaeological discovery that had ever been made in Japan. The site was occupied from around 5000 to 3500 BC. Covering an area of 35 hectares, this Jomon settlement is twice as large as any previously known Stone Age site in

Japan. To date approximately 100 post-built structures have been discovered in the central part of the site alone and are thought to have been used for the storage of food. Estimates place the total number of buildings at Sannai-maruyama at more than a thousand. These include hundreds of pit-houses as well as storehouses, watchtowers and other structures.

The remains of one particularly striking structure were discovered on rising ground at the perimeter of the settlement. Six large post holes were found in the ground, four of them still having pillars in place. The holes are 2.2 metres in diameter and 2.5 metres deep, whilst the pillars themselves – made from cedar trunks – are approximately 1 metre in diameter. Investigation of the surviving pillars of this building shows that they were deliberately placed to lean slightly inwards and that they were packed into the holes by alternate layers of compressed sand and clay. The depth of the holes and the care taken to secure the foundations of the building indicate that this was a high structure. The space between adjacent pillars is in each instance 4.2 metres, which may well indicate that some standard units of measurement were used by its builders. Its function and purpose remain unknown at present but it is thought to be a watchtower or perhaps even a religious shrine.

Archaeological research concerning this major site is still very much an ongoing project, and the full extent of the cultural attainments of its inhabitants will become gradually clearer as this work progresses. A truly enormous quantity of artefacts has already been unearthed from the excavated part of the site (only 15 per cent of the total area), so many in fact that they fill over 40,000 cardboard boxes! Among these are many items made from organic materials that rarely survive in sites of this age, including a variety of wooden tools (among them a number of paddles, one of which is 1.6 metres long), bark vessels, basketry and pieces of woven material. Three separate cemeteries have also been found at this settlement. Two of these burial areas were close to the main centre of the settlement and contain only the remains of children who were typically interred in large earthenware urns. So far about 900 such urns have been found at the site. The third cemetery contained only the remains of adults and was situated at the edge of the settlement. Whilst we can only speculate on the reasons for burying children closer to the living areas of the settlement than adults, the fact that the number of adult burials that have been found are little more than 10 per cent of the total suggests that the mortality rates of children in Jomon times were high.

There are a number of important features of the Sannai-maruyama site that demonstrate that Jomon culture was more complex than had previously been thought. Both the sheer size and the orderly arrangement of the settlement clearly indicate that it was carefully planned rather than simply a conglomeration of haphazard dwellings, rubbish dumps, burial grounds, storage areas and other buildings and constructions. This in itself shows that the community was well organised and that such planning could not have been undertaken without some kind of hierarchical organisation. It is also clear that far from being a temporary camp site of mobile hunters, this was a long-term sedentary village of considerable size. Not only were its inhabitants able to sustain their own economic needs through their activities in and around the settlement, but the great number of ceramic vessels and clay figurines found there strongly suggests that they may also have been supplying other, smaller Jomon communities in the vicinity with such products.

The diversity of the items produced at Sannai-maruyama also points to the existence of craft specialisation – in other words, certain sectors of the working community were involved in producing specific kinds of items for both the immediate community and surrounding ones. The presence of minerals such as obsidian and amber which are exotic to the region demonstrates that the circulation of goods on a local level was only one aspect of the settlement's economic interaction with the outside world. The aforesaid minerals had come from other parts of Japan and are clear evidence of long-distance trading networks. Although there is no evidence that rice was cultivated at Sannai-maruyama in Jomon times, there are indications that millet and other plants were perhaps cultivated by the Stone Age occupants of the site. Such evidence for a much more complex culture in the Jomon period than had previously been suspected has led Professor Yasuda Yoshinori of the International Research Centre for Japanese Studies to propose a radically new vision of what it means to be truly civilised. He has written:

> To truly be called a civilisation, a culture ought to have letters, a state, and metal tools. So I might be chided for using the term Jomon civilisation. People would say it's unacceptable, because even if the Sannai-maruyama ruins can be called a polis, the Jomon period still lacks other essential elements of civilisation.
>
> Now, however, the time has come to overthrow such outmoded concepts of civilisation and review the Jomon period based on a new concept of civilisation. If we adhere to the axiom that without a polis, letters, and metal tools the term civilisation cannot be applied, Jomon culture must forever be regarded as one in a primitive and barbarous stage of development. However, the society of the Jomon period had another marvellous principle that traditional civilisations have lacked: a respect for and co-existence with nature. The principle of living within the cycles of nature and maintaining social egalitarianism – nurtured for more than ten thousand years in the society of the Jomon period – is nothing less than what modern people yearn for in an era when the earth is endangered.
>
> The 'Discovery of Jomon Culture' means people must face the new environmental crisis and form a new idea of civilisation in order to survive. By insisting on the notion that civilisation must comprise a state,

letters, and metal tools, we make it impossible to find a way out of the crisis facing the modern industrial societies of the late 20th century. As I pointed out in *Jomon Culture in the History of the World*, 'Civilisation begins to appear when a workable system for living, that is, a proper relationship between man and nature, is established in accord with the features of a given region. This means that when an eternal and universal lifeway is formed that incorporates an ecological system indigenous to a region with the needs of its human inhabitants, civilisation begins to exist.'

Jomon culture, which established a permanent and universal way of life that was in harmony with the ocean and forests indigenous to the Japanese archipelago, is the true root of Japanese civilisation, and as such ought to be called Jomon civilisation.

Like Marija Gimbutas' advocacy of the essentially harmonious nature of the Old European civilisation before it was violently swept away, Yasuda Yoshinori's forceful statement concerning the Stone Age civilisation of Japan has elements of a blueprint for future social organisation as well as simply extolling the virtues of a bygone era. Historical civilisations have always drawn their inspiration from those cultures they consider exemplars worthy of being emulated. Modern Western civilisation has continually drawn inspiration from earlier civilisations, particularly that of Greece. Japan has looked to both Chinese civilisation and its own indigenous Yayoi period for its inspiration.

In both Japan and the West, the earlier Stone Age cultures represent a lowly and shameful ancestry which has been disguised and shunned in favour of the more respectable influence of historical civilisations. For a culture to deny its roots means that it will wither and die, and modern civilisation is, in its arrogant striving for progress at all costs, destroying the earth upon which its existence depends. Modern civilisation can be seen to be fuelled by the striving towards the controlling of an ever-increasing population by an ever-decreasing minority; seeking not only to dominate other cultures (to sustain itself it requires a massive underclass of economically subordinate nations) but to control nature itself, the latter at an expense of which we are now becoming increasingly aware through the prospect of ecological collapse. Cruelty and barbarism may even have increased under the reign of the historical civilisations, and it can be said with certainty that there is an increase in the social and technological means to inflict pain and suffering *en masse*. Tribal societies – the most savage of all modern peoples according to civilised opinion – are the very peoples that have been subjected to the most barbaric onslaughts of civilised nations. Colonial domination has given way to the apparently benevolent but equally paternalistic enterprise called development. The global attempt to 'develop' supposedly uncivilised peoples by imposing Western values, practices and 'know-how' has recently been referred to by anthropologist Mark Hobart as the 'growth of ignorance'. Western agencies with little or no knowledge of local ecological and social conditions implement agricultural and other projects which not only fail to deliver the goods but damage – sometimes irrevocably – the local ecosystem.

To sustain our own cultures we need to draw nutrition and inspiration from the roots, and if lost civilisations like those of Old Europe and Jomon Japan challenge our assumptions of progress then that is all to the good. Until we stop seeing prehistory as the longest and blackest of all Dark Ages, our understanding of the past, present and possible future of humankind will remain distorted. Like Gimbutas and Yoshinori, I believe that civilisations existed before the historical era and that our current view of what is and is not civilisation needs to be radically revised. The growing realisation that our current understanding of the meaning of civilisation is deficient may be perceived, as a sign of progress. As Leo Klejn has put it: 'it seems that the intellectual abilities of early man and the antiquity of human characteristics grow along with the growth of our own modern civilisation. Perhaps this means that our abilities to see and understand humanity are growing.' This growth of understanding about our prehistoric forebears is the subject of the rest of this book. The various innovations and achievements supposed partly or wholly to have developed in the civilisations of the historical era are shown to have been foreshadowed by the accomplishments of prehistoric cultures. Overwhelming evidence for these breakthroughs in the religious, artistic, scientific and technological spheres is brought together in the following chapters, showing not only that much that is presumed to have begun in historical times was in fact of Neolithic origin, but also that much considered to have arisen with the agricultural societies of even this time is, in similar fashion, shown to have developed much earlier in Upper Palaeolithic times. Finally, even some of the apparent innovations of behaviourally modern humans in the Upper Palaeolithic period are traced back to earlier times, to the Neanderthals and beyond, to the primordial origins of culture in the Lower Palaeolithic. Before embarking on this investigation of prehistoric religion, art, science, technology and even the origins of that most precious commodity of historical civilisation – namely writing – we shall first follow some remarkable lines of investigation by historical linguists who believe they can gain some access to the languages of the Stone Age.

Chapter 2

The Mother Tongue

It is estimated that there are between 5,000 and 10,000 different languages in the world (the discrepancy between these two figures is due to the different criteria used by various linguists in defining a language as more than a local variant or dialect of another), a fact which echoes the Biblical story of the Tower of Babel. But was there ever a single language in our remote past that preceded the confusion of tongues, and if so, is there any chance that we may be able to rediscover this Edenic state of communication? There have, of course, been attempts in modern times to create new languages, the most famous of these being Esperanto. However, such artificial attempts, noble as their intentions are, have had limited success and there seems no prospect that they will ever gain the universal currency that their creators had hoped for. Peoples throughout the world have always been aware that some of their neighbours spoke languages that were similar to their own whilst other cultures spoke entirely incomprehensible tongues. But the systematic attempts to bring some order to the worldwide confusion of tongues and classify the various languages into families that were related historically only began in earnest some two hundred years ago.

In 1786 Sir William Jones announced to the learned world that Sanskrit, the ancient language of India, was related to Latin and Greek, and so the Indo-European linguistic family was identified and shown to include a great number of distantly related relatives from as far apart as Ireland and India. The Iranian, Indian, Slavic, Celtic, Germanic, Italic, Greek, Lithuanian and Albanian are all branches of the Indo-European language group or family. Since Jones's highly important insight, historical linguists have been able to divide the world's languages into a number of families or groups. Bearing in mind the immensity of this task, it is hardly surprising that there has been considerable controversy amongst linguists concerning which languages and language groups are related. There are two basic approaches that have been taken by linguists. One approach is that of the so-called 'splitters', who are generally sceptical and cautious about proposing large-scale language groups consisting of more than one family (i.e. macro-families) on what they see as slender evidence. A bolder approach is taken by the 'lumpers', who are, as the name suggests, more inclined to see wider linguistic connections and, in some cases, claim to be able to reconstruct language families and their ancestral tongues way back into prehistoric times.

Those who have tried to place most (or, in some cases, even all) of the world's languages into macro-families do not always agree just how many such groups there are. The highly influential and controversial American linguist Joseph Greenberg and some other linguists are of the opinion that most of the thousands of human languages can be shown to belong to only 17 major linguistic groups. These 17 groups are listed below and their global distribution is shown on the map (*Figure 10*); estimated numbers of speakers of a language or group of languages are based on figures compiled by Merritt Ruhlen in 1987.

1. KHOISAN is a group of about 30 southern African languages with an estimated 120,000 speakers in total. Most of these are the languages of the so-called bushmen and Hottentots, who are distinct from the other peoples of Africa. It is thought that the ancestors of the Khoisan-speaking peoples may be the original inhabitants of this part of the continent. Two isolated languages of Tanzania (Hadza and Sandawe) are also placed in this group by some linguists. It is thought that such languages were once spoken throughout southern Africa and have declined as a result of the invasion of Bantu-speaking peoples and later of the Dutch colonialists.

2. NIGER-KORDOFANIAN is a very large group of African languages consisting of two branches. By far the largest branch is the Niger-Congo group (including all the Bantu languages such as Zulu, Rwanda, Kikuyu, etc.), which encompasses most of the southern half of the continent and is estimated to have about 180 million speakers. The smaller branch, the Kordofanian group, consists of approximately 30 languages spoken in Sudan.

3. NILO-SAHARAN languages are scattered across a number of African countries including Egypt, the Sudan, Mali and Tanzania. The group consists of about 140 languages with a total of 11 million speakers.

4. AFRO-ASIATIC is a group that includes the North African languages from Mauritania to Egypt and down to Somalia, as well as the Semitic languages of the Near East. Of the 240 or so languages that comprise the group, only a few are widely spoken. Arabic is by far the largest with some 100 million speakers (the total Afro-Asiatic group of speakers is about 180 million). Other well-known languages in this macro-family include Hebrew, Berber, Hausa and ancient languages such as Ancient Egyptian, Akkadian and Aramaic.

5. CAUCASIAN languages are spoken by about five million people in the Caucasus region, the group being made up of about 35 languages. The dominant language of the group, with over three million speakers, is Georgian. The group is divided into North Caucasian and South Caucasian (the latter is also known as Kartvelian).

6. INDO-EUROPEAN has already been discussed above. With some two billion speakers, nearly half the world's population speak one or

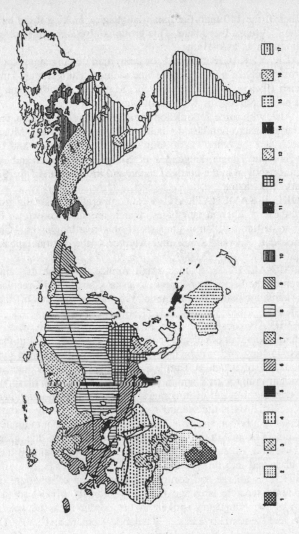

Figure 10

The principal language macro-families of the world (classified following Ruhlen 1987). The broken line indicates the appoximate northern limit of human dispersals during the Pleistocene period. (Arrows indicate the hypothetical Nostratic agricultural dispersals.) The language macro-families are as follows: 1. Khoisan; 2. Niger-Kordofanian; 3. Nilo-Saharan; 4. Afro-Asiatic; 5. Caucasian; 6. Indo-European; 7. Uralic-Yukaghir; 8. Altaic; 9. Chukchi-Kamchatkan; 10. Eskimo-Aleut; 11. Dravidian; 12. Sino-Tibetan; 13. Austric; 14. Indo-pacific; 15. Australian; 16. Na-Dene; 17. Amerind.

another of the 140 Indo-European languages, making it the largest of all the 17 groups listed here. This group is also sometimes referred to by linguists as Indo-Hittite.

7. URALIC-YUKAGHIR consists of more than 20 languages, spoken by approximately 22 million people, and includes Hungarian (14 million), Finnish (five million), Estonian, the Saami (or Lapp) languages, as well as the Samoyedic tongues of western Siberia.

8. ALTAIC, with some 250 million speakers and 60 languages, is one of the larger macro-families. It includes Japanese, Korean, Mongolian, Turkish and the other Turkic languages, such as Uzbek and Uighur, Manchu, the Tungus languages of Siberia and China and Ainu, a language spoken in the north of Japan and in the south of the Siberian island of Sakhalin.

9. CHUKCHI-KAMCHATKAN is a comparatively small group of north-eastern Siberian languages, which are also known as Palaeo-siberian or Luorawetlan. It consists of only five languages – Chukchi, Kamchadal, Koryak, Kerek and Alyutor – the various speakers of whom number only 23,000.

10. ESKIMO-ALEUT is another small family, of some 10 languages spoken by the Eskimos of Siberia, Alaska, Canada and Greenland and the Aleuts, the native inhabitants of the Aleutian Islands off the coast of Alaska.

11. DRAVIDIAN consists of over 20 languages, most of which are in southern India, but others are spoken in eastern India and in Pakistan. Approximately 145 million people speak these languages, the most well-known of which is Tamil. Some linguists have connected the Dravidian group with Elamite, an ancient language of Mesopotamia, thus creating a group of languages known as Elamo-Dravidian.

12. SINO-TIBETAN is the second biggest language family in the world with nearly a billion speakers, the vast majority of whom are Chinese speakers. The only other well-known languages in the group are Burmese and Tibetan; other minor languages of Nepal and India are also classified into this group.

13. AUSTRIC is a large and controversial amalgam of what are usually considered separate language families – namely Miao-Yao (a small group of four languages spoken in the border areas between south China and the northern areas of Thailand, Vietnam and Laos), Daic or Kadai (a group of about 60 languages spoken by roughly 50 million people, of which some 30 million speak the largest language of the group, namely Thai), Austroasiatic (consisting of some 150 languages spoken by about 56 million speakers; the group incorporates languages spoken in northern India and south-east Asia, the most well-known of which are Vietnamese and Khmer or Cambodian), and lastly Austronesian. Austronesian includes mainly island languages that range from Madagascar in the west, to the Philippines and Indonesia

and on into the islands of Melanesia and Polynesia. Its languages are spoken by about 180 million people, most of them in Indonesia and the Philippines.

14. INDO-PACIFIC is also known as Papuan and is a group of languages spoken on New Guinea and nearby islands. Despite the fact that there are only some estimated three million speakers for the entire group, there are over 700 Indo-Pacific languages. It is said that between 10 and 20 per cent of the world's languages are to be found in New Guinea.

15. AUSTRALIAN includes all the 170 languages of aboriginal Australia. Estimates place the total number of speakers at about 30,000. Although most linguists accept that all the Australian languages must be ultimately related, the connections between them are still, in many cases, obscure.

16. NA-DENE is a family made up of the Athapaskan languages of the native American peoples of Alaska and the western subarctic of Canada, their distant relatives the Navajo of New Mexico and Arizona, and three languages of the Pacific north-west coast of Canada and southern Alaska named Haida, Tlingit and Eyak.

17. AMERIND is a highly controversial macro-family proposed by Joseph Greenberg and consisting of all native American languages in both North and South America, with the exception of Na Dene languages and Eskimo-Aleut languages (both of which have been listed above). Approximately 600 Amerind languages are thought still to be spoken by an overall total of about 18 million people, and many linguists are unhappy to accept that they can all be put into such a large group, preferring to distinguish a number of distinct families without the overarching epithet of Amerind.

The classification of African languages into only four macro-families (Afro-Asiatic, Niger-Kordofanian, Nilo-Saharan and Khoisan) was the innovation of Joseph Greenberg, who published his research on this in the early 1960s. Although it was initially treated with considerable scepticism by the 'splitters', it has now been generally accepted. His subsequent study into the comparative linguistics of the New World led him to propose that all the indigenous languages of both North and South America could be grouped into only three macro-families. Two of the three proposed groups – Eskimo-Aleut and Na-Dene – were not really controversial at all and had already been identified by earlier linguists. The third macro-family, Amerind, was a different claim altogether, for not only did it break with the work of most earlier linguists, it was huge in comparison with Eskimo-Aleut and Na-Dene. Some of the 'splitters' in American Indian linguistics had identified about 200 separate language families in the entity Greenberg had given the name Amerind. This gives some idea of the vast gulf that divides the 'lumpers' from the 'splitters'.

There are some solid arguments in favour of Greenberg's three-family hypothesis for the New World. A dental study by Christy Turner of various Amerindian peoples showed that there were three distinct shapes of teeth in the native population of the New World, which indicated three separate prehistoric migrations from Asia to the Americas. Genetic studies of aboriginal Americans also indicate three distinct groups, which strengthens Greenberg's case further. According to most archaeologists, humans first entered the New World only 12,000 or 13,000 years ago. If this was the case, then the 'splitters' have to provide an explanation of how so many different language families arose in such a comparatively short time. Africa, which was populated at a vastly earlier date than the New World, is now, thanks to Greenberg, thought to have only four macro-families, so how can the Americas have 200 or so language families, as proposed by the 'splitters'? Whether Greenberg's American thesis will gain the widespread acceptance now enjoyed by his work on the language families of Africa remains to be seen, but unless the 'splitters' can give adequate answers to the genetic, dental and chronological problems that confront them, then he may well come out on top again.

Greenberg has also suggested that a macro-family named Eurasiatic encompasses the Indo-European, Uralic-Yukaghir, Altaic, Chuckchi-Kamchatkan and Eskimo-Aleut groupings and should be seen as the common ancestor of these languages, which were spread even in prehistoric times as far afield as Ireland at one extreme to the arctic regions of the New World at the other. He is not the only one or even the first to search so deeply for the roots of the linguistic tree. Right at the beginning of the twentieth century, the Danish linguist Holger Pedersen had expressed the opinion that there was a definite relationship between the supposedly distinct and independent language families of Indo-European, Semitic, Uralic, Altaic and even Eskimo-Aleut. On the basis of these links he believed that all these language groups were in fact descended from a remote language ancestral to them all which he called Nostratic, from the Latin *noster*, meaning 'our'. Pedersen's pioneering work on the Nostratic ancestral language has been built upon by various linguists, particularly Vladislav Illich-Svitych and Aron Dolgopolsky, who expanded the Nostratic macro-family to include yet more families (including Dravidian and Kartvelian or South Caucasian; it has also been suggested that Nilo-Saharan and Niger-Kordofanian should be added to the ever-growing extended Nostratic family), thus making it even larger than Greenberg's proposed Eurasiatic macro-family.

Languages that belong to the same family are shown to be related not only by their sharing of vocabulary but also by the similarities in both the structure and the sound of words in the various tongues. Because the various languages in a family are different, it is presumed that they are the offspring of a common ancestral tongue. Thus they can be seen as branches of a single language which is no longer spoken and can only be reconstructed from its descendants. The original language is known as a proto-language;

for example, the ancestor of all the Indo-European languages is called Proto-Indo-European, that of the Dravidian languages is called Proto-Dravidian and so on. Linguists have been able to estimate the age of proto-languages by comparing the various languages which make up the family in order to see the extent of diversity from the mother tongue. The geographical dispersal of the family members is also taken into account in assessing its antiquity. It is not possible to attain precise dates by this means, and even in the case of the well-known Indo-European family the experts cannot agree beyond saying that Proto-Indo-European should be dated to the Neolithic period, some time between 4000 and 6000 BC. According to the linguists who propose that there was a Nostratic ancestral language, Proto-Indo-European, Proto-Semitic, Proto-Uralic, Proto-Altaic and Proto-Eskimo-Aleut are all members of the Nostratic family, and it follows clearly from this that Proto-Nostratic must have been spoken at an earlier time than the various Nostratic languages! Dolgopolsky has described the sort of information that can be gained by the study of the vocabulary of the various proto-languages as 'linguistic palaeontology'. Such an approach can in some cases provide numerous clues into both the culture and the geographical location of the population that spoke the ancestral tongue, as he explains:

In Proto-Indo-European there are many words associated with agriculture and husbandry, suggesting that Proto-Indo-Europeans were a Neolithic people with a food-producing economy. On the other hand, among the 2,000 roots of the Proto-Nostratic lexical stock [i.e. Nostratic words that Dolgopolsky has reconstructed through the linguistic data of later languages] we do not find words suggesting acquaintance with agriculture or husbandry, but we do find many terms associated with hunting and food-gathering. If we take into account the fact that in other descendent proto-languages within the Nostratic family (Proto-Dravidian, Proto-Kartvelian, Proto-Altaic, Proto-Semitic, etc.) there is a rich terminology for agriculture and husbandry, we may suggest that Proto-Nostratic belongs to the period prior to the 'Neolithic revolution', while most of its descendent languages belong to the Neolithic epoch of food-producing economy. This is an important indication which may help us to suggest to historians and archaeologists where the Nostratic language might have been spoken. Another argument is the very location of the descendent languages: early Proto-Indo-European in Asia Minor, Proto-Hamito-Semitic in the Near East (and northeast Africa?), Proto-Dravidian probably somewhere in Persia, Proto-Kartvelian in Transcaucasia, Proto-Uralic and Proto-Altaic probably in Turkestan and adjacent regions. On the other hand, we know that the most ancient centre of Neolithic economy in western Eurasia was situated in southwest Asia. All this leads to a preliminary hypothesis that Proto-Nostratic was spoken in southwest Asia at a period prior to the 'Neolithic revolution', while most of its daughter-languages belong to the Neolithic epoch, and their spread over large territories of Eurasia and Africa was connected with the demographic explosion caused by the 'Neolithic revolution'. Of course, this is only a hypothesis suggested to archaeologists and proto-historians. Only by the joint efforts of archaeologists, proto-historians, and linguists, together possibly with anthropologists and geneticists, can we hope to achieve the truth.

The implications of the Nostratic hypothesis are mind-boggling. The theory proposes that most of the peoples of Europe and those in a large part of western Asia and parts of Africa were speaking Nostratic languages way back in prehistory, before the advent of agriculture. The reconstruction of Proto-Indo-European and the other Neolithic languages is remarkable enough, but the project of reconstructing the vocabulary of the Nostratic language takes us deeper into prehistory, back to the Upper Palaeolithic period, the latter part of the Old Stone Age! If the Nostratic language hypothesis is right, then it must be more than 10,000 years old and is likely to be nearer to 15,000 years old.

Another very archaic proto-language has been proposed which may be of a similar antiquity to Nostratic and equally, if not more, controversial. It does not have such a straightforward name and is simply called after the known linguistic groupings which are apparently descended from it. This is the Dene-Sino-Caucasian language proposed by Starostin and other linguists that includes languages as diverse as Basque, Chinese, Sumerian and Haida. If Dene-Sino-Caucasian is shown to be a genuine language group then it must, like Nostratic and Eurasiatic, be of Upper Palaeolithic age. The relationship between the Chinese language and the Na-Dene linguistic group was recognised in the 1920s by one of the great linguists of Amerindian studies, Edward Sapir. The connections between the North Caucasian languages and Sino-Tibetan are, in the opinion of the majority of linguists, tenuous at best. These large-scale reconstructions of the prehistoric ancestral languages of historical times are still at an early stage, but Nostratic, Eurasiatic, Amerind and Dene-Sino-Caucasian must be considered as a possible means of shedding light on the mysterious prehistory of the spoken word. It is remarkable that such a thing is possible at all, and as Vitalij Shevoroshkin has said, William Jones's insight into the links between Sanskrit and European languages was revolutionary at the time in a way comparable to the status of Nostratic and other linguistic reconstructions today.

As speculative as the ideas of the great macro-families of languages such as Amerind, Dene-Sino-Caucasian and Eurasiatic are, some linguists, most conspicuously Merritt Ruhlen, believe that they can identify correspondences between even these vast groups and are seeking to reconstruct the primordial ancestor of all the world's languages, a language called either Proto-Global or Proto-World. Ruhlen and his colleague John D. Bengtson have put forward some 45 global etymologies which they believe indicate a connection between all of the world's language families. Some of these, as they readily acknowledge, had been proposed earlier by Greenberg and other linguists. Just how striking the correspondences that they have collected are can be shown by a few examples. Each of these etymologies has been given a gloss which characterises the 'most general meaning and phonological shape of each root'. Thus these etymologies incorporate a correlation in respect not only of the meaning of the words but also of their sound. The seven global etymologies given below detail the basic terms for man, woman, child, hole, vulva, finger and water. In their linguistic investigations Bengtson and Ruhlen came across correspondences for these terms in a hundred or more different languages belonging to many of the 17 language families detailed above. For the sake of brevity I have simply provided a few examples as a representative sample of their findings.

MANO meaning 'man' appears in various language groups in Afro-Asiatic (Ancient Egyptian *Min*, the name of a phallic god; in Somali *mun*, 'male'), Nilo-Saharan (e.g. in Tama, an East Sudanic language, *ma*, 'male'), Dravidian (e.g. Tamil *mantar*, 'people, men', Gondi *manja*, 'man, person'), Austric (the Yao people whose language belongs to the Miao-Yao group call themselves *man* or *mun*), Amerind (in the Squamish language of Canada the word *man* means 'husband', whilst in South American Indian languages there are numerous correspondences, such as the Wanana word *meno*, 'man', and, in the same language, *manino*, 'her husband'; Kaliana *mino*, 'man, person', *imone*, 'father-in-law', Guahibo *amona*, 'husband') and Indo-European (including the English word *man*).

KUNA meaning 'woman' is found in Afro-Asiatic (e.g. in the Cushitic language Oromo in which *qena* means 'lady'), Indo-Pacific (in the Bea language of the Andaman Islands in the Indian Ocean *chana*, 'woman', Tasmanian *quani*, 'wife, woman'), Altaic (e.g. in the Central Asian Turkic language Kirghiz *kunu*, 'wife'), Australian (Gamilaraay *gunijarr*, 'mother'), Amerind (Guarani *kuña*, 'female', Kamayura *kunja*, 'woman', Suya *kuña*, 'woman', Cuica *kunakunam*, 'woman') and Indo-European (for example the English word *queen*).

MAKO meaning child appears in various forms in Dravidian (Tamil *maka*, 'child', Telugu *maga*, 'male'), Sino-Tibetan (Burmese *(sa-)mak*, 'son-in-law'), Indo-Pacific (as in the south-west New Guinea language Jaqai *mak*, 'child'), Amerind (Zuñi *maki*, 'young woman', Waikina *make*, 'son', Coto *ma-make*, 'boy') and Indo-European (in Old Irish *macc*, 'son', Old English *magu*, 'child, son, man', and Swedish *måg*, 'son-in-law').

K'OLO meaning 'hole' has been traced in the Khoisan language group (*kxolo*, 'nostrils' i.e. nose hole), Nilo-Saharan (as in the Saharan language

Kanuri *kuli*, 'anus', East Sudanic language Nandi *kulkul*, 'armpit'), Uralic-Yukaghir (Finnish *kolo*, 'hole, crack', Hungarian *halok*, 'incision', Zyrian *kolas*, 'crack'), Altaic (Korean *kul*, 'cave', Japanese *kur*, 'hollow, scoop out'), Dravidian (Tamil *akkul*, 'armpit'), Sino-Tibetan (West Tibetan *kor*, 'hollow in the ground, pit'), Austric (Tagalog *kilikili*, 'armpit') and Indo-European (English *hole*).

PUTI meaning 'vulva' has been shown to exist in Niger-Kordofanian (as in the Malinke word *butu*, 'vulva'), Nilo-Saharan (Gao *buti*, 'vulva'), Afro-Asiatic (Hebrew *pot*, 'vulva', in the Chadic language Jegu *paate*, 'vulva'), Uralic-Yukaghir (Ostyak *puti*, 'rectum'), Dravidian (Tulu *puti*, 'vulva'), Austric (Ami *puki*, 'vulva'), Amerind (Yamana *puta*, 'hole', Guahibo *petu*, 'vagina', Jaricuna *poita*, 'vagina') and Indo-European (Old French *pute* – modern *putain* – 'whore').

TIK meaning 'finger' or 'one' appears across the globe as in Niger-Kordofanian (Gur *dike*, '1'), Nilo-Saharan (Dinka *tok*, 'one'), Afro-Asiatic (Hausa *(daya)tak*, 'only 1'), Altaic (Korean-*teki*, '1', Old Korean *tek*, '10', Japanese *te*, 'hand', Turkish *tek*, 'only'), Eskimo-Aleut (Greenlandic Eskimo *tik(-iq)*, 'index finger', Aleut *tik(-laq)*, 'middle finger'), Na-Dene (Tlingit *tek*, '1'), Amerind (Karok *tik*, 'finger, hand', Mangue *tike*, '1', Katembri *tika*, 'toe'), Indo-Pacific (in the Boven Mbian language of south-west New Guinea *tek*, 'fingernail') and Indo-European (Latin *dig(-itus)*, 'finger', *decem*, '10').

AQ'WA meaning 'water' is found in Nilo-Saharan (e.g. in Nyimang *kwe*, 'water', and Kwama *uuku*, 'water'), Afro-Asiatic (Janjero *ak(k)a*, 'water'), Altaic (Japanese *aka*, 'bilge water', Ainu *wakka*, 'water'), Amerind (Allentiac *aka*, 'water', Culino *yaku*, 'water' and *waka*, 'river', Koraveka *ako*, 'drink', Fulnio *waka*, 'lake') and Indo-European (Latin *aqua*, 'water').

It seems absolutely remarkable that such correspondences should exist across time and space and that languages found as far afield as the deserts of southern Africa, the Amazon rain forest, the Arctic and the cities of Europe still retain links from a remote time when they were all closely connected. Of course the meanings and sounds of such words are by no means strictly identical across the globe, but are we able to provide a better explanation of these striking correspondences than that provided by Ruhlen and Bengtson – namely that they all derive from a common mother tongue? There are only two other possible explanations for the similarities between these words from diverse language families. Firstly it could be suggested that the words were borrowed from one language family to another. Borrowing is certainly something that sometimes occurs between two languages but can hardly provide anything like an adequate explanation for the numerous links that clearly exist between language families across the world. The other possible alternative explanation is that these correspondences are merely accidental and that if a determined linguist ploughs through the dictionaries and word lists of the thousands of different languages, then he or she is bound to come up with some plausible-sounding correspondences. Yet repeated accidental resemblance of both meaning and sound on a global scale is too unlikely to contemplate.

That such parallels exist between language groups in distant parts of the world is striking and is hard to dismiss simply as mere coincidence, but the idea that a Proto-Global language could be recovered even in part is simply too much for many linguists to take and as such remains highly controversial. Even most proponents of the macro-families are sceptical of this kind of research, some believing it to be impossible to substantiate at all, whilst others feel that it is simply premature to go this far back without sufficient knowledge to reconstruct even the macro-families. Colin Renfrew, Disney Professor of Archaeology at the University of Cambridge, has pointed out that if it were shown that Proto-Amerind and Proto-Eurasiatic were in fact connected – a possibility that he does not reject outright but is highly sceptical of – this would take us back over 20,000 years ago to some time before these two macro-families must have split to go their separate ways. Ruhlen is more than aware of the implications and is apparently undaunted about the time scale involved in the reconstruction of Proto-Global for he says:

What if Bengtson and I are right, and the linguistic similarities we have uncovered really do represent traces of a single earlier language family? What would be the implications for archaeology – and human prehistory – of this fact? First of all, it must indicate that the origin of *present* linguistic diversity is fairly recent, otherwise similar words in many different families would simply not be found. But what is 'fairly recent'? If Indo-European can be distinguished by anyone on the basis of a dozen words – and no one seems to doubt this – and if Indo-European is 6,000 to 8,000 years old, depending on whose views of the origin of Indo-European one subscribes to – is it then too much to expect that trained linguists, through a careful sifting of all the world's linguistic evidence, might not be able to extend the linguistic horizon four or five times beyond the obvious? What seems to me the most probable explanation for the linguistic data, as they are presently known, is that current linguistic diversity derives from the appearance of behaviourally modern people forty or fifty thousand years ago. While anatomically modern humans may have appeared in Africa before 100,000 BP, these people did not behave like us. That in itself may indicate their more rudimentary linguistic skills. Several scholars have in fact suggested that the 'sapiens explosion', as it is sometimes called, involved the development of fully modern human language as recently as 40,000 years ago (or perhaps somewhat before).

In saying this Ruhlen is implying that he is on the trail of Proto-Global, a language, or group of languages, that may possibly date back to 40,000 BP, the beginning of the Upper Palaeolithic, that is to say the time when the so-called Human Revolution took place. To many people this may seem to be a fantastic theory too neat and way, way too early to be convincing. The macro-families proposed by Dolgopolsky, Starostin and Greenberg are still highly disputed by many linguists, some of whom believe that it is simply impossible for historical linguistics to say anything of value – let alone certainty – about the state of any language whatsoever before about 4000 BP. Ruhlen, in adding an extra nought into the equation, has gone far beyond the limits which many linguists have set themselves and indeed others in their field. Clearly, only further researches into both the macro-families and the Proto-Global hypothesis of a single mother tongue will establish whether such exciting and bold researches gain general acceptance. Such developments in the field of historical linguistics are based on the idea that it is possible to reconstruct, at least in part, prehistoric language groups from which the languages known from historical times are believed to be derived.

Although it would seem obvious that this kind of research should be of profound interest to prehistorians, the archaeological community have been rather slow in their attempts to assimilate this mine of important new information. This may be partly due to the fact that many archaeologists believe that they have their work cut out in trying to unravel the complex and still mysterious origins of comparatively late and completely uncontroversial language groups such as Indo-European. Trying to locate the geographical and chronological origins of the Indo-Europeans and relating them to the archaeological records of Europe and Asia has been anything but straightforward. In fact there is still no general consensus among archaeologists as to the homeland of the Indo-Europeans nor even the dating of their migrations. If such a well-established linguistic group presents so many problems for the archaeologist, then how can he or she hope to grapple successfully with the more elusive and archaic entities like Nostratic? Nevertheless, there are archaeologists who feel that simply to ignore the dramatic developments in historical linguistics on the grounds that such new information makes the task of understanding prehistory even more complicated is a mistake. The most prominent advocate of this more positive approach is Colin Renfrew. Renfrew is clearly aware of the problems with the Indo-Europeans, as his own theory of their origin and dispersion is one of a number of the conflicting ideas on this subject. Despite this he has advanced some tentative ideas on how the macro-families of languages that existed in prehistory could shed light on the archaeological record.

Renfrew suggests that the various proto-languages that are said to belong to the Nostratic group could have dispersed from the zone in which agriculture seems to have first developed, namely in the Near East and Anatolia (modern-day Turkey). In this scenario the expansion of these languages beyond the region would be directly associated with the spread of farming. The parent language, Proto-Nostratic would thus be located somewhere in the core region and obviously to a time preceding the origins of agriculture. Renfrew proposes a date of around 15,000 BP for this ancestral language. He considered and rejected the idea that Nostratic could be traced back to the Gravettian period of the Upper Palaeolithic. The Gravettian period, around 25,000 years ago, is distinguished by both the distinctive technology of its stone tool-kits and also by its art, which is epitomised by the discovery of numerous female figurines of a broadly similar style found across Europe from France to the Ukraine. Both the tool technologies and these artistic objects (commonly known as Venus figurines; for further information on the importance of these objects, see Chapter 14) are sufficiently distinct for archaeologists to identify the Gravettian cultural complex to be an extremely widespread phenomenon that suggests that communication must have taken place between western and eastern Europe at this time. It is for this reason that some shared language group can be suggested for the Gravettian cultural complex. Renfrew rejected the idea that Nostratic languages might have been spoken among the various Gravettian communities on a number of grounds, including the fact that the Gravettian period was simply too early for the apparent emergence of Nostratic around 10,000 years later.

If any of the proposed macro-families of languages might be plausibly linked to the Gravettian, then Renfrew felt it was something akin to the Dene-Sino-Caucasian reconstruction, or Proto-Dene-Caucasian as he calls it. In this very tentative reconstruction of events, Renfrew suggests that such languages might have been spoken in various parts of Europe and northern and central Asia and then subsequently been overlain by the Nostratic languages emanating from the Near East and Anatolia. He suggests that the early language of the Sumerians (which is not one of the languages included in the Nostratic macro-family) may date back to the Upper Palaeolithic period and have provided the spoken background to what was later to be written down as the first written script some 5,000 years ago. In the next chapter we will see how it was not only that the spoken language of the Sumerians preceded its expression in written form but that the script itself grew out of prehistoric roots. Despite the profound importance of research which claims to be able partially to reconstruct earlier language families than those known since historical times, archaeologists have, in the main, taken little notice of such developments into the prehistory of linguistics. Colin Renfrew has been the leading figure in bringing these linguistic findings into the sphere of archaeology and has shown that to ignore them is to miss out on a completely new tool for the prehistorian to use.

Chapter 3

A New Rosetta Stone

It is generally agreed that the earliest known writing comes from the Ancient Near East and first occurred during the period 3500 to 2800 BC. Expert opinions favour 3100 BC as the most likely time for this major historical landmark to have taken place. It was in the city of Uruk (in present-day southern Iraq), epicentre of the Sumerian world, that this great innovation took place. This initial form of writing has been called proto-cuneiform, as it precedes the cuneiform script, which derives its name from the Latin *cuneus*, 'wedge', *forma*, 'form', on account of the wedge-like impressions made on the clay tablets which embody these earliest written texts. It was followed shortly afterwards by the Proto-Elamite script in south-western Iran and, a little later, by Egyptian hieroglyphics. About a thousand years later, writing appeared for the first time in the Indus Valley civilisation. The Sumerian innovation of writing is considered to have spread to Iran, Egypt and probably the Indus Valley through the trade networks known to have existed during these times. The emergence of scripts in both China and later in Mesoamerica is thought to be independent of outside influences and thus they represent entirely distinct traditions.

The three main prerequisites for civilisation – city dwelling, the formation of capital and the use of writing – are usually said to have first come together dramatically at Uruk. Decades of excavations at Uruk by German archaeologists have unearthed a wealth of fascinating evidence on the development of writing which undermines traditional theories as to how the first script came into being. During the eighteenth century, when people were becoming dissatisfied with the then prevalent notion that writing had been delivered to mankind by God or the gods, new theories arose that attempted to explain the genesis of scripts in evolutionary terms. The most influential of such theories, which, in modified form, still has its supporters today, was that of William Warburton. He maintained that the abstract characters that make up scripts had developed out of an earlier, cruder 'picture writing', citing Egyptian hieroglyphics as a prime example.

Archaic tablets found at Uruk during the late 1920s and early 1930s not only revealed the great antiquity of Sumerian writing but also contained evidence that was eventually to deal a death blow to the pictographic theory for the origin of writing – at least in the realm of the Mesopotamian scripts.

The earliest texts from Uruk were written in a script consisting of abstract characters which bore little or no resemblance to the 'picture writing' they should have been. As the old theory faded, a new prototype was sought to explain the emergence of writing. The complexity of the earliest script makes the idea that writing, and thus civilisation as commonly understood, arose, as it were, almost overnight a plain absurdity. Whilst some scholars supposed an even earlier and as yet unknown script to be the source of Sumerian, Elamite and Indus Valley scripts – thereby begging the question – one researcher came across the answer whilst looking for something else.

The answer to this great enigma did not come about through the discovery of some hitherto undeciphered text but rather through the realisation by the archaeologist Denise Schmandt-Besserat that thousands of apparently insignificant and rather mundane clay objects (which she dubbed 'tokens') that have been found throughout Ancient Near Eastern archaeological sites, far from being insignificant or mundane were the basis of a system that had been completely overlooked. In her major work *Before Writing* she has described her initial realisation that she was on to something as follows:

I must say that the tokens came my way by chance. It all started in 1969–1971, when I was awarded a fellowship from the Radcliffe Institute, Cambridge, Massachusetts, to study the use of clay before pottery in the Near East. This led me to visit systematically Near Eastern archaeological clay collections dating from 8,000 to 6,000 BC stored in museums of the Near East, North Africa, Europe, and North America. I was looking for bits of Neolithic clay floors, hearth lining, and granaries, for bricks, beads, and figurines, and I found these aplenty. I also came across a category of artifacts that I did not expect – miniature cones, spheres, disks, tetrahedrons, cylinders, and other geometric shapes. The artifacts were made of clay and belonged, therefore, to my study. I noted their shape, color, manufacture, and all possible characteristics, counted them, measured them, sketched them, and they entered my files under the heading 'geometric objects'. Later, the term *token* was substituted when it became obvious that all the artifacts were not in geometric form; some were in the shape of animals, vessels, tools, and other commodities.

I became increasingly puzzled by the tokens because, wherever I would go, be it Iraq, Iran, Syria, Turkey, or Israel, they were always present among the early clay assemblages. If they were so widely used, I reasoned, they must have had a useful function. I noted that the tokens were often manufactured with care and that they were the first clay objects to have been hardened by fire. The fact that people went to such efforts for their preparation further suggested to me that they were of importance. I sensed that the tokens were part of a system because I repeatedly found small and large cones, thin and thick disks, small and large spheres, and even fractions of spheres, such as half and three-quarter spheres. But what were they for?

When she quizzed other archaeologists she drew a blank. No one knew what they were for. Schmandt-Besserat had realised that she was on to something and was hooked. Delving deeper into the literature on the archaeology of the region, she found an article describing a hollow, egg-shaped tablet found at the site of Nuzi in northern Iraq and dated to the second millennium BC (see *Plate VIII*) The cuneiform inscription that was written on it says:

> Counters representing small cattle:
> 21 ewes that lamb
> 6 female lambs
> 8 full-grown male sheep
> 4 male lambs
> 6 she-goats that kid
> 1 he-goat
> 3 female kids
> The seal of Ziqarru, the shepherd.

When the archaeologists who had found the tablet opened it up they discovered 49 counters, exactly the total number of animals listed on the outside. Clearly this was some sort of accounting system using counters (or tokens). To Schmandt-Besserat this was a revelation, and she described this hollow tablet and its contents as 'the Rosetta stone of the token system'. It was now becoming clear to her that this find shed light on the prehistoric tokens she had become preoccupied with. The study of the tokens became her main work and over the next 15 years she assembled a vast body of convincing evidence (based on analysis of over 10,000 tokens and related artefacts) to show that a highly effective system of accounting had existed since the early Neolithic period of the Near East (8000 BC). But that was not all. This accounting system was also the prehistoric forerunner of both the archaic texts of Uruk and the development of a written numerical system. As will be seen below, her account of the developmental stages from the first Neolithic tokens to the inception of writing is brilliantly and convincingly argued.

The tokens in question range between 1 and 5 centimetres across, are almost all of clay, and are made into distinct shapes, each of the sixteen categories having a particular role in the overall system as the means of counting different commodities. The early tokens, dating from about 8000 to 4400–4300 BC, are called plain tokens and mainly consist of a few types – spheres, cones, tetrahedrons, discs and cylinders. In the second phase of the system, beginning around 4400 BC, complex tokens appear (and, for the first time, include tokens of all sixteen types), most significantly in the activities

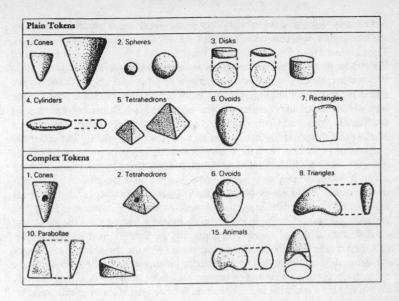

Figure 11

and administration of temples during the fourth millennium BC (see *Figure 11* for examples of both the simple and complex types of tokens).

Thus the earliest types of tokens (plain tokens) were not replaced in the second phase, but were rather augmented by the complex tokens. Exactly where the token system first developed is not, as yet, clear. Among the earliest sites to yield tokens are level III of the Syrian site Tell Mureybet (8000 BC), level E of Ganj Dareh in Iran (also about 8000 BC) and Tepe Asiab (*c.* 7900–7700 BC). Although Tepe Asiab is assigned to the earliest phase of the Neolithic period, the indications are that it was a site occupied by people who still lived mainly by hunting and gathering. Nevertheless, despite their economy still displaying such features of the Mesolithic period, their use of clay not just for tokens but also to fashion figurines indicates the transition was well underway.

What is curious is that of all token-bearing sites of the time (not just in Iran but also elsewhere in the Near East), Tepe Asiab boasts the greatest overall number (193) and variety of types of tokens (9). This shows that from the very beginning of the tokens in the archaeological record the system was already fairly sophisticated. At the later site of Jarmo in northern Iraq, which seems to be an average farming community of the seventh to sixth millennium BC, about 2,000 tokens have been discovered, more than at any other site of the period. This does not mean that this was some sort of centre for the manufacture or distribution of the tokens but rather that tokens have simply been recorded there systematically by archaeologists, whereas elsewhere they have gone unnoticed or unrecorded. The scale of token finds from Jarmo may thus be a fairly typical example of the extent of token use in comparable communities of the time.

So what exactly were these tokens for, what did the different types represent and why did they quite suddenly arrive on the cultural scene in the Neolithic period? The tokens were, according to Schmandt-Besserat, a completely new way of dealing with information. They represent a 'conceptual leap' from earlier means of encoding data. The essential purpose of the tokens from the start was to keeps records of numerous kinds of items. In order to make it clear which type of thing was represented by a token it was necessary to create a whole system consisting of distinct forms of token that could easily be identified. Thus spherical tokens, cylindrical tokens and all the other initial forms were explicitly fashioned to designate distinct groups of things. For example, judging from proto-historical and early historical use of comparable signs by the Sumerians, the conical type of token most probably stood for a measure of grain, the sphere for a larger measure of grain, the ovoid represented a jar of oil, the cylinder a domestic animal, the tetrahedron a unit of labour, and so on. Tokens of different kinds and in different quantities (e.g. one cone for each measure of grain) could be stored together. Such accounting records precluded the need for any individual to memorise the accounts, as they could be referred to and understood at a future date by anyone who knew what the various token types represented. The token system was open-ended in the sense that if there was a requirement to account for new types of goods a new kind of token (either of a new shape or an existing shape augmented with new and distinctive markings) could simply be added to the existing ones without disrupting the system. Schmandt-Besserat suggests that the use of tokens involved a basic level of syntax. She says that:

> It is likely, for example, that the counters were lined up on the accountant's table in a hierarchical order, starting on the right with tokens representing the largest units. Such was the way the Sumerians organized signs on a tablet, and it is logical to assume that the procedure was inherited from a former [i.e. Neolithic] way of handling tokens.

Despite the considerable capacities of the token system it was easy to use, and because it did not rely on a spoken language for its workings it was easy for those speaking different tongues to adopt. This seems to be exactly what happened, judging from the extremely wide distribution of tokens through-out the Near East during Neolithic times. Because one ovoid stood for one jar of oil, two ovoids stood for two jars of oil, three ovoids for three jars of oil and so on, it was, of course, easy to add or subtract items using the system. But when the numbers of things to be counted were very high the ease and efficiency of the system turned into a tedious and time-consuming business. Although certain token types were clearly developed to deal with this problem (such as the large tetrahedron which may well have represented a week's work for one man), it was, to an extent, a consequence of the basic principle of one-to-one correspondence upon which the whole system was built. Also the need to create more and more token types to deal with the number of items that required accounting eventually led to the system collapsing in on itself and a new system of handling data had to be developed; in short, the complex tokens became too complex as a system and this precipitated the final steps towards the emergence of the Sumerian script.

Before describing this next stage in the process we need to know what kind of economic activity was behind the simple token system. Although it seems that there was a widespread uniformity in the system that straddled numerous communities, the system itself was not a means to record long-distance trade transactions. Whilst there is abundant evidence for the existence of such trading networks, the tokens do not appear to be implicated in recording such exchanges. Their *raison d'être* appears to lie with the social changes that occurred with the development of Neolithic cultures. Simply put, whilst the hunter-gatherers of the Palaeolithic lived in basically egalitarian communities, Neolithic societies developed an increasing level of hierarchy. In order to administer the accumulation of foodstuffs and other goods that proliferated in the era of early farming, local élites arose. The village headman would be required to collect and then redistribute the economic bounty (or paucity) of the community. And it is here that the tokens come in. The token system was the means by which the flow of goods could be accounted for and the means by which the necessary calculations and decisions of such a redistributive economy could be made to work efficiently. As the communities of the Near East grew in size and diversified their activities, there was a need to expand on the simple tokens that had served their users admirably for 4,000 years.

The complex tokens that arose to augment their simple counterparts during the fourth millennium BC are most conspicuous in two major cities – Susa, the main city of Elam, and Uruk itself. At Susa 783 tokens have been found, an amount which far exceeds other finds at both contemporary and earlier Iranian sites. In contrast to the finds from Jarmo (which, whilst large, cannot be used to show its pre-eminence in the token world), at Susa the sheer number of both tokens and, even more tellingly, of 190 distinctive sub-types, indicates that the proliferation of diverse forms of goods and the need to account for them was a pressing matter for the metropolitan accountants of the time. At Uruk, 812 tokens have been recovered and reveal 241 sub-types. Thus the total number of tokens and the numerous sub-types from the two great cities are roughly equal, and there are other parallels too. The tokens are also so similar in size, colour, markings and the quality of clay used that even Schmandt-Besserat – who has surely scrutinised more tokens than anybody since the time of Uruk! – says that they are indistinguishable. This huge expansion of the token system was a response to the need to account for the goods produced in the craft workshops of the great cities. In Uruk, Susa and other Near Eastern urban centres of the fourth millennium BC, the great majority of tokens were found in public rather than private buildings, indicating that the flowering of the complex token system was both developed and administered by the temple bureaucrats of the time. Schmandt-Besserat states that:

> The southern Mesopotamian temple appears as the final outcome of a four-thousand-year-old economy of redistribution. It inherited from the past a tradition of pooling communal resources and still relied on the continuous outpouring of goods in kind delivered as gifts to the gods. The emergence of the city-state brought, however, a major transformation to the age-old redistributive system by establishing taxation; that is, the obligation for all individuals or guilds to deliver a fixed amount of goods in kind under penalty of sanctions.

Among the innovations that facilitated the preservation of these accounting methods were two novel ways of storing tokens. In the first procedure tokens were perforated and tied together with string. This device is found with many complex tokens and does not appear to have been generally used in collections which include only simple tokens. The second procedure, developed around 3700–3500 BC, involved the use of clay envelopes to put the tokens in (a link that Schmandt-Besserat had first made when she came across the Nuzi hollow tablet, her 'Rosetta stone'). The further innovation of the accountants of Uruk was to impress markings in the shape of the enclosed tokens on the outside of the envelope and then seal it. The latter of these procedures was the crucial step towards the development of the Sumerian script.

The two-dimensional impressions that were added to the outside of the envelope may have begun as a way of making doubly sure that the accounting was accurate and checks on the tokens could be done without breaking the seal and removing them from the envelope, but it was not to remain so. At a certain stage the accountants realised that the notation on the outside of the envelope made the tokens and the envelope itself redundant. Thus the old system was replaced by a simpler and more efficient method that was used during the period 3500–3100 BC. The signs that were previously ancillary to the tokens replaced them and the envelope gave way to the tablet. Accounting could now proceed by simply impressing the symbols for the things to be recorded directly on to a tablet, thus moving the whole system one crucial step closer to the development of a script. The evidence that this was in fact the case is found in the mutations that took place between the stage of tokens with notations on their envelopes and the impressed tablets proper. Schmandt-Besserat describes two such instances:

At Susa, a solid ball with flattened surfaces impressed with signs can be considered a transitional stage between envelopes and tablets. The subsequent changes consisted in the flattening of the impressed tablets. Falkenstein noted, however, that the earliest pictographic Uruk tablets were sometimes strikingly convex, perhaps still perpetuating the roundish shape of the former envelopes.

Her revolutionary theory was also corroborated by the similarities in the way information was recorded on both sets of artefacts. The tablets not only used the same kind of seals that were used on the envelopes (both cylinder and stamp seals) but also repeated the same sets of signs and organised them in the same way, namely in parallel and horizontal lines. This shift from complex tokens contained in impressed envelopes to tablets was not a response to changing economic conditions, as far as we can tell. Whilst plain tokens appear to have arisen to serve the needs of the ranked societies of the early Neolithic and complex tokens to accommodate the rise of urban-based industrial activities, no comparable economic shift accompanies the first appearance of pictographic tablets. What we see is rather the accounting system running on its own steam, as it were, towards a further level of abstraction. With the use of pictography at Uruk came the recognition, around 3100 BC, that quantities previously always referred to only in concrete instances (three jars of oil, three measures of grain, etc.) could be abstracted and expressed by symbols (such as 'three') that could then be applied to any given instance. That is to say, they invented numerals. The subsequent development of phonetic writing belongs to the realm of history and, as such, takes us beyond the period with which we are concerned.

Schmandt-Besserat's discovery of the significance of the prehistoric token system has not only brought to light a hitherto unknown aspect of Neolithic life, it has also provided an eloquent and convincing account of the subsequent chain of events that led to the world's first known script. She has shown that writing did not develop overnight but was the outcome of activities over a period of thousands of years. Yet to avoid making the same kind of error that has sometimes been made for the origin of writing – namely that it sprang up out of nowhere – one is tempted to suggest that this Neolithic token system (which even at its inception shows the use of different types of token to count different kinds of things) might have some direct precursor from the Mesolithic period. The almost simultaneous appearance of tokens in different parts of the Near East in the eighth millennium BC seems to indicate just such a possibility. Could it be that some – as yet unknown – local Mesolithic traditions of record-keeping were simply adapted to the new medium of clay and so became the tokens that Schmandt-Besserat discovered? She rejects such a possibility as no such evidence has been forthcoming from Near Eastern Mesolithic sites. She argues that because the early Neolithic tokens of the various regions of the

Near East are standardised in their forms (cones, discs, cylinders, tetrahedrons, etc.), it is far more likely that the invention took place in one particular region at the outset of the Neolithic and spread fairly rapidly from there. The present state of archaeological knowledge makes it impossible to ascertain where this original centre for the manufacture and use of the tokens might have been. She also points out that the shapes of the plain tokens indicate their innovative character; being, as they are, distinctively formed from clay, the Neolithic material *par excellence* 'are very easy to achieve; they are in fact the shapes which emerge spontaneously when doodling with clay'.

Whilst she is undoubtedly right to see the token system *per se* as a Neolithic invention, it will be shown when we delve even deeper into the prehistoric records that the Mesolithic, and even Palaeolithic, roots of Neolithic systems are more complex and tangled than she would allow. In the opinion of Schmandt-Besserat the means of encoding and storing information during the Palaeolithic and Mesolithic periods were limited essentially to a very basic level. The most common form of devices used at this time were tallies in the form of notched bone or antler which, whilst able to store data, had certain 'built-in' limitations. According to her:

> This first stage in data processing signified two remarkable contributions. First, the tallies departed from the use of ritual symbols by dealing with concrete data. They translated perceptible physical phenomena, such as the successive phases of the moon, rather than evoking intangible aspects of a cosmology. Second, they removed the data from their context. For example, the sighting of the moon was abstracted from any simultaneous events such as atmospheric or social conditions. Finally, they separated the knowledge from the knower, presenting data . . . in a 'cold' and static form, rather than the 'hot' and flexible oral medium, which involves voice modulation and body gestures. As a result [such tallies] not only brought about a new way of recording, handling, and communicating data but an unprecedented objectivity in dealing with information.
>
> The tallies remained, however, a rudimentary device. First, the notches were nonspecific since they could have an unlimited choice of interpretations . . . in fact, the notched bones were limited to storing only quantitative information concerning things known by the tallier but remaining enigmatic to anyone else. These quantities were entered according to the basic principle of one-to-one correspondence, which consisted of matching each unit of a group to be tallied with one notch. Finally, because tallies used *mostly* [my italics] a single kind of marking, namely notches, they were confined to handling only one kind of data at a time. One bone could keep track of one item, but a second one was necessary to keep track of a second set of data.

Schmandt-Besserat's view that the prehistoric tallies first arose *after* human beings were already making use of ritual symbols and had developed a cosmology is actually a loaded statement. Tallies and other intentional markings on bones and other materials were made right from the onset of the Upper Palaeolithic, i.e. from the Aurignacian age onwards. Thus for ritual symbolism and cosmological ideas to have existed before this time would place them back in the Middle Palaeolithic period, to the time of the Neanderthals. Even today there are many archaeologists who would deny the Neanderthals such capacities, although as will become apparent there are indications of both ritual burial (see Chapter 16) and of cosmological symbols (see Chapter 6) as early as the Middle Palaeolithic.

Furthermore, as will be described in Chapters 5 and 6, the Upper Palaeolithic notation systems were by no means as limited as she suggests. She concedes that in addition to the tallies it is likely that Palaeolithic and Mesolithic peoples made use of naturally occurring objects such as stones or twigs as counters, as have many pre-literate societies in historical times. Yet even in such a usage she sees the same limitations she has ascribed to the Palaeolithic tallies, namely the inability to keep track of more than one category of item at a time. But there is no reason not to think that more than one category of objects could be recorded at one time by using, say, sticks to represent one type of item, pebbles to represent another and so on. Furthermore, there are ethnographic examples of pre-literate and non-agricultural peoples who have used rather complex methods of counting. The Neolithic system of tokens may thus be innovative in its material (clay), its repertoire of distinctive shapes (tetrahedrons, cylinders and the rest) and even its purpose (to account for the goods of the new kind of economy – cultivated cereals and domestic livestock), but nevertheless *not* represent an evolutionary leap in cognition over the Palaeolithic systems of notation. But before we explore the notation systems of the Mesolithic and Palaeolithic periods it is necessary to consider other developments in the Neolithic period that have little in common with the Near Eastern token system with its emphasis on accounting procedures.

Chapter 4

The Signs of Old Europe:
Writing or Pre-Writing?

Prehistoric pottery shards incised with signs were discovered in the 1870s at the archaeological site of Turdas (spelt Tordos in the earlier accounts) in Cluj, Transylvania, in western Romania. Similar signs have been found on various artefacts from other sites belonging to the Vinca culture, which was named after the type site of Vinca, near Belgrade. These signs have long attracted the attention of scholars seeking to trace cultural links between the Balkans and the more 'civilised' regions of the eastern Mediterranean, the Levant and Mesopotamia. At the beginning of the twentieth century, Arthur Evans (the son of John Evans, an important archaeologist in his own right), who almost single-handedly revealed the splendours of ancient Cretan culture to the world, made mention of the Turdas signs in connection with his discussions of the emergence of written scripts. Other scholars of the time also showed some interest in comparing these signs with those found in Troy, Egypt and elsewhere. Partly due to the influence of the great authority on later prehistoric Europe, V. Gordon Childe (1892–1957), Vinca culture was to become perceived as an outlying cultural entity essentially influenced by external, more 'civilised' forces, and he paid the Turdas and other Vinca signs scant attention. The earlier interest that had been shown in this enormous corpus of signs faded, particularly in western European archaeological circles, and was not revived until an important discovery in the 1960s.

In 1961, Dr N. Vlassa, an archaeologist working at the Historical Museum of Cluj, was excavating the Transylvanian site of Tartaria (some 20 kilometres from Turdas), which had not been investigated since war-time excavations that had been undertaken during 1942 and 1943. Vlassa found that there were four distinct cultural layers at this prehistoric site. Although most of the artefacts that he found were of an unremarkable nature, his discovery of three clay tablets (see *Figure 12*) was to send a shock wave through the archaeological world. The tablets became, for a while, the focal point for the fierce debates of the time that were raging over the chronology of prehistoric Europe. Radiocarbon dating was, at the time, a comparatively new method of dating archaeological remains, and the early dates that those using the technique came up with for south-east Europe were way earlier

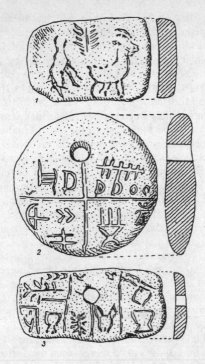

Figure 12

than those on which conventional archaeological wisdom was based. In fact, the signs on the tablets were to play a vital role in the debate concerning what has been called the radiocarbon revolution.

In the lowest (and therefore earliest) layer – belonging to the Vinca culture – Vlassa discovered a sacrificial pit filled with 'ashy earth'. At the bottom of the pit were found 26 burnt clay figurines, two alabaster figurines, a bracelet made of spondylus shell and the three tablets. Close by this heap of artefacts were the broken and burnt bones of a single individual estimated to be between 35 and 40 years old. Vlassa interpreted these finds as the remains of a sacrificial rite and, like many other archaeologists, jumped to the conclusion that some sort of ritual cannibalism had taken place. As he provided no supporting evidence for cannibalism we can treat this lurid suggestion with a healthy degree of scepticism.

All three of the tablets are made of sandy clay and are reddish-brown in colour. The first tablet, roughly rectangular in shape, is 5.2 × 3.5 × 1.6 centimetres and is interpreted by Vlassa as a hunting scene. There are three separate images. Two of them are fairly clear; on the right is a depiction of a goat and in the upper part of the centre of the tablet is an ear of corn. The third image, on the left, is more obscure and has been interpreted as a rearing animal or perhaps a human figure. The other two bear a number of signs, arranged in rows, that may suggest a form of script. Of these two, one is a disc-shaped tablet, pierced by a round hole, and 6.1 × 6 × 2.1 centimetres in size. The other is roughly rectangular in shape and also has a round hole. This tablet measures 6.2 × 3 × 0.9 centimetres.

Vlassa, suspecting that the signs on the Tartaria tablets may represent a form of writing, immediately began to speculate that, if they were, then a Near Eastern origin must be sought, since the idea of prehistoric Europeans developing writing on their own was considered too unlikely a possibility to take seriously. Archaeological theorists – such as V. Gordon Childe – saw the Near East as the source of almost all cultural developments of any significance, and so the appearance of such a highly developed grouping of signs on the tablets must be due to the direct or indirect influence of oriental 'high' cultures; in short, of Mesopotamian civilisation. Vlassa saw the archaic tablets of Uruk as the closest analogies to the signs on the tablets from Transylvania. Following this line of reasoning, which seemed perfectly sensible at the time, he suggested that since the Uruk clay tablets dated from the period 3500–3200 BC, the Tartaria tablets would appear to date from somewhere between 2900 and 2600 BC, a time lag that was sufficient for the Mesopotamian innovation to have reached Transylvania. This dating fitted rather neatly with the dating of south-eastern European prehistoric developments that had been developed without the aid of radiocarbon dating. Vlassa kept an open mind, being well aware that if the radiocarbon dates for the region were shown to be correct then the Tartaria tablets would be far earlier than their counterparts in Mesopotamia and would date them back to about 4000 BC, and thus almost a millennium before the Mesopotamian invention of writing.

The German scholar Adam Falkenstein, who was the great pioneer in the analysis of the archaic writing systems of the Sumerians and who published the first of the texts from Uruk in the 1930s, put his considerable academic weight behind Vlassa's opinion that the two Tartaria tablets bearing signs (numbers 2 and 3) closely resembled Mesopotamian tablets, particularly those from Uruk III B (which belong to a period before the emergence of cuneiform). Falkenstein viewed the first tablet (Vlassa's 'hunting scene') as belonging to a style derivative of Mesopotamian seal impressions. Sinclair Hood, who had been the Director of the British School of Archaeology in Athens from 1954 until 1962, entered the debate in 1967. He was also very forthright in confirming the apparent similarity between the Transylvanian and Mesopotamian tablets, and wrote in the leading British archaeological journal *Antiquity* that:

> The signs on the Tartaria tablets, especially those on the roundel no. 2 [the disc-shaped tablet], are so comparable with those on the early tablets from Uruk ... as to make it virtually certain that they are somehow

connected with them. Several of the signs appear to be derived from Mesopotamian signs for numerals. The only difference is that on the Mesopotamian tablets the whole shape of the sign in the case of numerals was sunk in the clay with a round-ended stylus, while at Tartaria the equivalent signs were incised in outline. In addition the shapes of the tablets and the system of dividing groups of signs by means of incised lines recur in Mesopotamia. The Mesopotamian tablets are normally convex on one side or on both, but some larger tablets are relatively flat like nos. 1 and 3 at Tartaria. There are also a few small circular tablets among those from Uruk to compare with no. 2. But the string-holes on two of the Tartaria tablets appear to be a feature without any parallels among the early tablets of Mesopotamia.

Whilst accepting the analogies with Sumeria, Hood noted that some of the signs of the Tartaria tablets were also similar to those found in the Minoan scripts of Crete, although they were clearly not Cretan. The string-holes, whilst absent in Mesopotamia, were a common feature of Cretan tablets. Despite the parallels, he was of the opinion that it by no means followed that the signs on the Tartaria tablets represented an advanced stage of writing. In seeking to explain the connections with Mesopotamia, Hood suggested that Sumerian prospectors, drawn particularly by the great gold-bearing deposits of the Transylvanian region, may have been instrumental in establishing cultural connections between the two regions. The considerable distance between them was not seen as a problem by Hood, who noted that the Indus Valley script (which is generally accepted to have been influenced by Mesopotamian writing systems) developed in a region as far from Mesopotamia as is Transylvania. He suggested that if the Tartaria signs were shown not to be a genuine script, then their ultimate Mesopotamian origin could be due to some kind of Near Eastern 'missionary' activity in which the mysterious signs from the east were eagerly adopted by the magician-priests of prehistoric Europe, who, whilst ignorant of their meaning and proper use, nevertheless imitated their forms for their magical power. The Hungarian scholar Janos Makkay stated in no uncertain terms that 'the Tartaria pictographs' ... Mesopotamian origin is beyond doubt' and echoed Hood's opinion that the tablets represented a garbled and 'senseless' mimicry of Near Eastern writing by someone who had either seen the 'real thing' in the Near East or observed the use of tablets by Sumerian merchants visiting Transylvania.

The idea that the barbaric priests of Europe sought to imitate their cultural 'superiors' from the east by superstitiously seizing upon writing for magical purposes is of little value in unravelling this complex situation. The ancient civilisations of both Mesopotamia and Egypt were saturated with magical thought, as their own respective written traditions clearly show. Literate civilisation and a belief in magic are thus by no means incompatible. As I have shown in the introduction to this book, magic has played a prominent role throughout human history and cannot be easily dismissed as a form of primitive thought. Since the symbols contained in the Tartaria tablets are earlier than Near Eastern writing, Hood's suggestion becomes a red herring anyway. The Transylvanian symbols and their counterparts elsewhere in the Old European world were clearly part of a magical tradition that owed nothing to the Near East.

As I have already said, the Tartaria tablets became embroiled in the arguments concerning the incompatibility between traditional and radiocarbon dating for prehistoric south-eastern Europe. Adherents of the more conservative dating used the discovery of the tablets in order to demonstrate what they considered to be the unreliability of radiocarbon chronology. Their argument ran as follows. Many of the signs found on two of the Tartaria tablets (and the iconography of the other one) show clear and undeniable parallels with Mesopotamian writing and therefore must be later than the emergence of the Sumerian script. It follows from this that radiocarbon dates that assign sites like Tartaria to prehistoric periods way before the rise of Near Eastern civilisation are wrong. This line of reasoning grew more and more untenable as it became obvious that, in the main, the radiocarbon dates were correct. Clutching at straws, some of the advocates of the traditional chronology even suggested that Vlassa had been mistaken in assigning the tablets to the lowest cultural layer at the site and that they actually were artefacts belonging to the Bronze Age. Despite these last-ditch attempts to hold back the tide of change, the radiocarbon revolution succeeded in overthrowing the traditional chronology for prehistoric Europe. Once this new order was established, the only way in which the Transylvanian-Mesopotamian connection could be maintained was to argue that it was not the Sumerians who were the innovators, but rather that their 'barbaric' counterparts in Europe had somehow influenced them.

The Tartaria tablets are by no means the only artefacts of their type to have been found in south-eastern Europe. A number of other objects that are of comparable age and also display a similarly complex use of signs have been unearthed. The Gradesnica 'plaque', as it is usually called, discovered at the site of Vratsa, near Gradesnica in western Bulgaria in 1969, has been assigned to the Chalcolithic period and is between 6,000 and 7,000 years old. It is not really a plaque as such but seems rather to be a shallow vessel that has been inscribed on both its inner and outer surfaces (see *Figure 13*). The top left of the illustration shows the outer side, which depicts a human-like figure with its arms raised in a ritualistic posture, surrounded by a number of signs. The inner side (bottom left) is divided by four horizontal lines with what appear to be three signs on each line (also shown on the right). It has been suggested that the vertical lines that make up some of the signs on the inner side may represent numbers. Signs on both sides of the vessel are analogous to ones found elsewhere and belonging to the Vinca cultural region. A disc-shaped stamp seal unearthed at the central Bulgarian

Figure 13

site of Karanovo (and known as the Karanovo seal) is probably at least 5,500 years old and also displays a number of similar features to the Tartaria tablets. On the Karanovo seal, the signs are divided into four groups by the arms of a cross, which divides the surface of the disc into four quadrants. Although the early date and the content of such artefacts have occasionally been put forward to bolster theories that the Sumerians and other 'advanced' cultures derived their scripts from the Balkans, this is unlikely to have been the case.

We have already seen that Schmandt-Besserat has convincingly shown that the Sumerian script could quite naturally have developed from its home-grown economic roots; namely from a system of Near Eastern Neolithic accounting. And anyway the idea that the invention of writing could be assigned to Europe rather than Asia is too far-fetched for most scholars to countenance. With the acceptance of the new radiocarbon chronology, there was only one other explanation available. Since the Tartaria tablets were earlier than Sumerian writing, they could not be real writing and their apparent resemblance was simply coincidental. It was on that note that, once again, the Vinca sign system faded into comparative obscurity, at least as far as most mainstream archaeologists were concerned.

Looking in hindsight at the debate concerning the cultural source and significance of the signs on the Tartaria tablets, several points may strike the reader as rather strange. How can a number of highly respected experts such as Falkenstein put their considerable academic weight behind the argument *for* the Tartaria tablets being a kind of script clearly deriving at least some of its elements from Mesopotamian writing (whether or not the Transylvanian copyists understood its true purpose or not) only to have their views utterly shattered by the incontrovertible radiocarbon dates? Would the attempt to link Transylvanian 'writing' with that of Mesopotamia ever have been attempted had scholars of the time known that the Tartaria tablets were earlier rather than later than their Sumerian counterparts? And, if the tablets do not represent a knowledge of writing nor even a 'primitive' attempt to imitate the activities of Near Eastern civilisations, then what is the meaning of these signs?

Since the links that have been proposed between the Vinca sign system and the development of writing in Mesopotamia have been shown to be spurious, it seems more productive, at least for the moment, to treat the two traditions as basically distinct, each being developments in their own right. As Schmandt-Besserat has shown the largely indigenous development of Mesopotamian script, so others have been working on the Vinca signs as an innovation of south-eastern Europe dating back to at least 4000 BC (although the dating of the prehistoric cultures of the region is still the subject of some disagreement and some archaeologists view the signs as developing as early as 5500 BC). Whilst a number of mainly eastern European scholars have brought forward strong arguments against seeking Near Eastern prototypes for the European 'scripts', few attempts have been made to analyse systematically the corpus of Vinca signs and to ascertain what their purpose and meaning might have been. Two particular interpretations of the signs are of considerable interest. The charismatic Lithuanian archaeologist Marija Gimbutas, whose controversial theories concerning the goddess religion of prehistoric Europe have made her widely influential far beyond the confines of the archaeological establishment, developed a reading of the signs which she expanded in the latter part of her life to a much wider geographical and chronological framework than that of 'Old Europe', the term by which she designated the Neolithic and Chalcolithic cultures of south-eastern Europe. Before considering her translation of the 'Language of the Goddess', as she called it, the more conservative but nevertheless highly important study of one of another scholar named Shan Winn provides the link between previous work on the Vinca signs and Gimbutas' own far-reaching reconstruction of the spiritual life of ancient Europe.

During the 1970s, Winn compiled the largest catalogue of Vinca signs to date and in the process incorporated hundreds of examples of the use of the signs found on artefacts in museums that had never been published before.

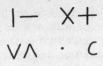

Figure 14

In order to make some sort of sense of the disarray that confronted him, Winn ordered the apparently chaotic nature of the signs using a number of interlocking classifications. He analysed the signs according to which kind of artefacts they appeared on (amulets, altars, figurines, pottery, spindle whorls, etc.), whether they appeared in isolation or as part of a group of signs, their relative frequency in different areas and different times, and so on. Having examined thousands of occurrences of the signs, he was able to discover that the corpus consisted of 210 signs. Most of these signs are composed of straight lines and are rectilinear in shape, and only a minority incorporate curved lines, a fact which may be due to the technological restrictions of incising them on to clay rather than any cultural preference for straight rather than curved types of signs. All the Vinca signs were found to be constructed out of five core signs. These elemental building blocks of the entire system are: (1) a straight line, (2) two lines that intersect at the centre, (3) two lines that intersect at one end, (4) a dot and (5) a curved line (see *Figure 14*).

The various ways in which these basic five elements could be repeated or combined led Winn to develop 18 categories of signs. As these 18 categories are basically a means developed by Winn to analyse their frequency on different kinds of artefacts and so reveal the symbolic patterns of the Vinca culture, I shall not burden the reader by describing all these categories in detail, but will instead simply give one or two instances of signs that belong to one of the categories. For example, Winn's sixth category consists of various signs that are composed out of core sign number 2 (two lines that intersect at the centre) plus secondary elements. Signs that belong to this category include a cross with a single straight line attached to only one of the cross's four arms; the swastika; and an X with repeated V-shapes projecting from the upper two arms. By dividing the Vinca signs into these various categories, Winn is not suggesting that the prehistoric users of them also divided them in the same way; it is rather a means by which he is able to create some artificial order which can, in turn, lead to more direct insights into the original use of the signs. Using such methods, Winn was able to establish some important conclusions about the Vinca signs. It became clear that far from being used in an arbitrary fashion, the signs were used systematically by their makers and a number of clear patterns emerged. As Winn puts it:

Internal analysis of the Vinca signs supports the conclusion that the signs are conventionalised and standardised, and that they represent a corpus of signs known and used over a wide area for several centuries. One set of signs is confined to figurines, whilst several signs are especially evident on figurines but are occasionally used on other artifacts. Signs on pottery are distinguished as a whole; when complex signs are employed on pottery, they occur only once, and then only as isolated signs, never in sign groups.

Certain frequent signs occur only as isolated signs, whereas others occur only in groups, and an important set of signs occur both as isolated signs and as components of sign groups. Signs which occur only in sign groups are virtually excluded from use on pottery, while signs found only in an isolated context are almost exclusively pottery signs. Important sign clusters are distinguished by a significant association with a particular type of object; i.e. specific signs are inscribed on pottery and other signs on figurines. [Spindle] whorls play an intermediate role, having both a distinct nature and sharing components pertaining to pottery; this duality may be reflected in their roles in both domestic and religious affairs.

The Vinca signs thus represent a sophisticated system of communication that was undoubtedly of great cultural importance for its users. Nevertheless, according to Winn it would be misleading to describe it as writing, since even the most complex series of signs do not make 'texts'; they are simply too short and there are too few repetitions of signs for them to be considered as script comparable to that of, for example, the Sumerians. Yet the clear and distinctive use of different signs in different contexts has led Winn to describe the Vinca as a system of pre-writing. In comparing the signs which are important only in isolation with those that frequently occur in groups, a significant contrast comes to light. It is the signs that appear in groups that have the simpler shape, and many of them – unlike the Vinca signs commonly found depicted on their own – are reminiscent of those signs that are found in written scripts elsewhere.

Thus there are different levels of complexity in the Vinca system. The isolated signs found mainly on pottery probably represent the lowest level of Vinca communication through signs and belong to the household sphere. The spindle whorls, whilst displaying a more sophisticated use of signs than pottery, rank below the highest levels of communication that seem to be embodied in the Tartaria tablets, the Gradesnica plaque, the Karanovo seal and other comparable finds. These latter kinds of artefacts have no obvious utilitarian function and it would appear that they had a symbolic and ritual purpose that removed them from the humdrum concerns of daily life. Also, the fact that they display the most complex use of the signs strongly suggests an important ceremonial value, and thus their association with a priestly élite who would have officiated in religious practices. Unlike the early Sumerian tablets, or for that matter the Neolithic token system that preceded them, the various levels of the Vinca system do not suggest that they were of much economic significance. The Vinca signs seem to derive from religious rather than material concerns, and the longer groups of signs may be some kind of magical formulae.

Having demonstrated that the search for a Mesopotamian origin for the Vinca signs was a vain quest, we must ask if there are any more fitting comparisons or even direct links between the prehistoric signs from the Balkans and signs from other regions. Such comparisons can be made with marks on pottery and other artefacts that appear in the Aegean and Mediterranean areas, particularly those found in Egypt, Troy (in what is now westernmost Turkey) and the ancient city of Phylakopi on the island of Melos in the Aegean. What is even more surprising than the fact that there is a high correlation between the Vinca and other signs is the fact that the most extensive use of signs (in both the range of artefacts decorated with them and the occurrence of groups of signs) took place in the Balkan region and not in the more 'civilised' areas to the south.

The distinguished Egyptologist Sir Flinders Petrie (1853–1942) made an extended study of Predynastic and later Egyptian occurrences of such signs and made it quite clear that rather than these early marks being the origin of Egyptian writing they were, in fact, a separate system that existed before and then later alongside the hieroglyphs. Petrie was also aware of the similarities between the Egyptian signs and those found elsewhere in the Mediterranean region, and suggested that they may have been some kind of international *lingua franca*. He also expressed the belief that because of their similarity of form with the signs that were later used in alphabetical scripts, these early signs may well have something to do with the origins of the alphabet. Other early investigators also expressed the opinion that the various groups of signs in the Aegean and Mediterranean cultures might well represent an early, even prehistoric, forerunner of writing. I shall return to this fascinating avenue of research in the next chapter. Before that I wish to pick up another thread of the story and return to the enigma of the signs of Old Europe, which, as will become clear, is interwoven with these wider issues.

The social order that is said to have existed in Old Europe has already been described in Chapter 1, and therefore I shall deal only with Marija Gimbutas' reading of the signs from this period of European prehistory. Her eloquent arguments for a radical revaluation of currently dominant ideas about what civilisation actually means include a stronger view of the indigenous European development of writing than that of Shan Winn. Winn could only bring himself to describe the Vinca signs as pre-writing, but for Gimbutas, and for others such as Harald Haarmann of the University of Helsinki, they are the real thing. As we know, this was not an entirely new assertion, but most of those who had previously characterised the Tartaria tablets and analogous Vinca signs as genuine writing did so on the mistaken assumption that they were later than Sumerian and could always be neatly 'explained' as somewhat pale imitations of Near Eastern intellectual innovations. We have also seen how many scholars, on realising that the Vinca signs were simply too early to be derived from Mesopotamia, abruptly dropped the question of the European script. Without the comfort of leaning on Near Eastern origins, few of them dared to pursue the question of an independent emergence of writing in Europe. For others, who had tried and failed to bolster the traditional chronology for prehistoric south-eastern Europe by invoking the Tartaria tablets as a refutation of radiocarbon dates, the tablets were simply dismissed as meaningless jumbles of signs. Thanks to the efforts of Winn, Gimbutas and Haarmann, it is hard to see how the whole question can be stifled again, and it seems destined at last to be given the serious attention it deserves.

Gimbutas sees two distinct periods in the development of the Old European script, an Early Phase (*c.* 6000–5300 BC) and a Climactic Phase (*c.* 5300–4000 BC). Haarmann traces the objects that are inscribed with the script from the late sixth millennium to their decline and eventual disappearance around 3500 BC which he, in line with Gimbutas, attributes to the overthrow of the Old European civilisation in the Danube valley by the Indo-European hordes. With an ever-increasing domination of the continent by the Indo-Europeans, the indigenous populations of south-eastern Europe were faced with two choices: they could either remain where they were and so be ruled by their new masters; or they could migrate in search of new lands in which they could continue, albeit in a modified form, the Old European lifestyle. It appears that they did both. Among those who chose to flee their ancient homelands were groups that found a haven to the south, on the shores of the Aegean and beyond. Haarmann describes the diaspora as follows:

> There are clear indications that Old European patterns survived on the periphery. The rise of Cycladic culture, the earliest evidence of which dates back to about 3200 BC, is definitely related to the migration of pre-Indo-Europeans, and Crete is also reached by a wave of new settlers during the course of the third millennium BC. Neolithic culture in Crete had been based upon a foundation which is common to South-eastern Europe and which was reinforced by the migration of pre-Indo-European refugees to the Aegean islands.

Haarmann notes that there are a number of striking parallels between the various strands of the pre-Indo-European cultural fabric – especially those related to religious symbolism and mythology. Among these common features is the use of the bull and the snake as important religious symbols. In the case of the snake it is a form of the goddess intimately intertwined with the bird goddess motif in both Old European and the later Cretan iconography. The bee and the butterfly are also recurrent divine attributes, and the butterfly is represented by the distinctive motif of the double axe. Haarmann sees the goddess mythology of Old Europe echoed in these motifs that also feature prominently in the ancient civilisation of Crete. He then traces the links between the Old European script – as found in the Vinca culture – and later systems of writing, particularly those of Crete. We may again draw attention to the contrast between the emergence of script in Mesopotamia and in Europe. As was made clear in an earlier chapter, the economic nature of the Sumerian script was an outgrowth of a long tradition of accounting stretching back to the beginning of the Neolithic in the region. In Old Europe the main concern that seems to have led to the use of a script was religious rather than material, suggesting a discontinuity of cultural emphasis, at least in terms of communicating through a system of signs. It must not be presumed that the Old European script can be dismissed on the grounds that it does not follow the Sumerian pattern of emerging out of economic requirements.

There are other instances of the development of writing for magical purposes. Although the origins of writing in China are still rather mysterious (some scholars seek to explain it as ultimately deriving from another cultural centre such as the Near East; others see it as an indigenous development), what is beyond dispute is that the oracle bones, inscriptions from the Shang dynasty, dating back to the second millennium BC, are examples of writing concerned not with economics but with the practice of divination. Before returning to the debate concerning the Old European script, it is noteworthy that some Chinese scholars have sought precursors of these inscriptions on prehistoric pottery signs, and bearing in mind the Neolithic influences on the Sumerian script, this may be an approach that will lead to the resolution of the problem of the origin of writing in China.

The actual form of the signs which make up the Old European script is basically linear, and this led Gimbutas to describe the script as 'Old European Linear signs'. When she did so she clearly had in mind the Cretan writing system known as Linear A. Whether Linear A is an Indo-European, Semitic or 'Old European' language is as yet unknown; Haarmann explains its undeciphered nature as supporting his theory that it is the latter. According to Haarmann, the stereotypical view of the development of Cretan writing – namely that it came about because of the economic requirements of palace administration – is a mistaken one. Here again we can see the tenacity of the Sumerian model for the origin of writing. Haarmann counters this by asserting that it was rather the medium for priestly activities of a practical nature in the case of Linear A, and of a more ceremonial nature in the case of the Cretan hieroglyphic script. Clearly the linear nature of Old European and Cretan Linear A, the apparent religious emphasis of both and the proposed migrations of populations from the ancient homeland to Crete are not sufficient grounds to prove that the two are linked. There are, however, a considerable number of parallels in the inventories of the two systems of signs (see *Figure 15*). Haarmann has found some 50 signs that are common to the two and believes this is far too many to be explained away as a coincidence. He is, nevertheless, cautious and does not jump to unwarranted conclusions:

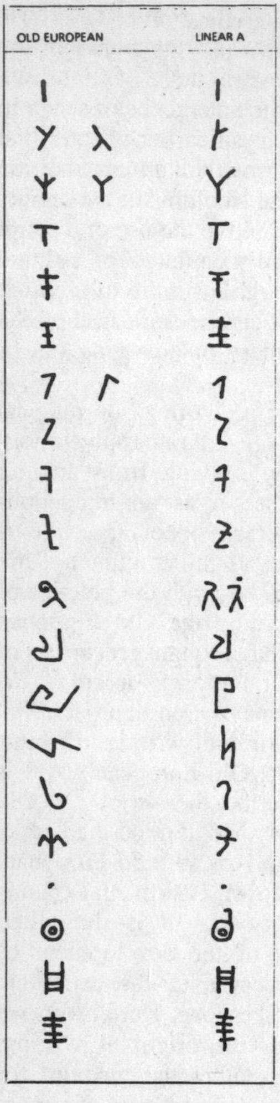

Figure 15

It would be a misconception to say that the Cretan Linear A is a derivation from the Old European script. However, almost one third of the sign inventory of the Old European script has been revived in the Linear A sign list where these elements make up almost half of the total inventory. Cultural continuity in this regard is equal to a fragmentary maintenance of original symbols which retained their original function as signs of a writing system in the culture ... which adopted them, rather than to the transfer of a complete writing system from one cultural complex to another.

The central problem in understanding the cultural role of the non-Indo-European scripts in any detail is simply that no one, at least as yet, can read them. Even the comparatively late Cretan writings in Linear A and hieroglyphic scripts will not yield up their secrets. Until another Rosetta stone may be found with a bilingual text containing a translation into an Indo-European language, decipherment remains only a dream. The vastly older Old European script presents far greater difficulties; even if one or more of the later non-Indo-European scripts were to be deciphered, the problem would remain. If the Old European script is a highly developed form of writing – and Gimbutas and Haarmann have presented a credible case for it being so – its very antiquity makes it a book that seems to be destined to remain firmly shut.

There are, perhaps, those who will be relieved to see proponents of an Old European script stuck, as it were, on the historical side of what appears to be an impassable barrier. If the script cannot be deciphered, then it will always be possible to dismiss it. But the open-minded will find its parallels to Cretan script and its own internal order and sophistication arguments sufficient not to reject it out of hand. The notion of an Old European script goes against many of the entrenched positions of archaeology and the traditional view of the development of civilisation. The implications are immense. Its general acceptance as a form of writing would lead to a number of basic assumptions being brought into question. It would mean rejecting the notion that the 'high' civilisations (using the word in its more traditional and 'acceptable' sense) of the Near East some 5,000 years ago did not invent writing but were preceded in this breakthrough far earlier and, worse than that, by 'barbaric' Stone Age Europeans. Even more fundamentally, the beginning of history (heralded by the use of a script and thus the production of written records) would be cast back into the 'primitive' darkness of Stone Age prehistory. The ideological wall constructed to divide prehistory and history, the primitive and the civilised, and writing and pre-writing would fall overnight were the Old European script to be indisputably vindicated. It would herald nothing less than the collapse of the present notion of civilisation. But the quest for the roots of writing, which have already been detected in the Neolithic periods of both the Near East and Europe, leads us further back in time to even more remote periods of prehistory.

Chapter 5

The Palaeolithic Origins of Writing

In seeking to explain the emergence of the Old European script, Gimbutas proposed that it was part of a much wider corpus of signs that expressed the cosmological and spiritual beliefs of the Neolithic age. These symbolic designs and motifs (crosses, spirals, dots, lozenges, etc.) appear on numerous artefacts made from bone, stone, wood and clay. Many of the surviving objects from the period, such as cult figurines, ritual vessels, lamps, spindle whorls and stamp seals, are embellished with elements of this symbolic language. Gimbutas believed that such designs were not merely decorative but were elements of 'an alphabet of the metaphysical' and as such can provide us with a key to understanding the religious life of the Stone Age. In 1989 she described her quest to decipher these symbols as follows:

> Some twenty years ago when I first started to question the meaning of the signs and design patterns that appeared repeatedly on the cult objects and painted pottery of Neolithic Europe, they struck me as being pieces of a gigantic jigsaw puzzle – two-thirds of which was missing. As I worked at its completion, the main themes of the Old European ideology emerged, primarily through analysis of the symbols and images and discovery of their intrinsic order. They represent the grammar and syntax of a kind of meta-language by which the entire constellation of meanings is transmitted. They reveal the basic world-view of Old European (pre-Indo-European) culture.

The interplay of such signs and their organisation into numerous combinations were the means by which the Old European cultures expressed their metaphysical notions. Ideas concerning time, space and the nature of the cosmos were expressed through a set of signs. Clearly if this symbolic language is to be 'read', then each component sign must have its own distinct place in the overall system. Gimbutas believed that her research revealed the inner workings of this system. The Old European system, whilst expressing beliefs about the world, life and death, time and space and other universal concerns of humanity, did so through the lens of Neolithic culture and its particular preoccupations.

For example, the symbol of the cross and associated derivative symbols such as the swastika embody a number of the principles of this prehistoric ideology. The cross represents the four corners of the world and was a symbol of the cosmic or yearly cycle. For the Neolithic farming communities, this was naturally associated with the practices of agriculture and the birth and death of vegetation. The cross is an ideogram for wholeness, cyclical time and the renewal of life. Gimbutas shows that it is a symbol particularly associated with the Great Goddess of life and death and the goddess of vegetation. Another sign in the system, the spiral, symbolises the universal snake, which in turn is the embodiment of the dynamism, life force and regenerative powers of nature. The dynamic energy represented by the snake image and the spiral seems to be intimately associated with the life-giving element of water. In Old European iconography one of the most important members of the pantheon is the snake goddess. She is represented either as part human, part snake, or simply as a human figure covered with spiral motifs. In the Neolithic period as in the ancient world, the symbol of the serpent is also associated with rainfall – an aspect of nature that is obviously of fundamental concern to all farmers.

Initially Gimbutas deciphered the meaning of the cross, the spiral and many other signs from her investigation of the thousands of cult objects that have been unearthed from south-eastern Europe. In her later work she discovered that it was possible to find clear parallels to these symbolic traditions in archaeological remains from other parts of Neolithic Europe, and extended the term Old Europe to incorporate them. In fact, she came to equate the terms Old European and pre-Indo-European. She also began to delve deeper in order to establish which of the signs might date back to earlier periods of the Stone Age in Europe, to the Mesolithic and Palaeolithic. The study of such symbols shows that as the ancient world drew inspiration from the Neolithic, so too were the earliest farming communities of this period drawing inspiration from the symbolic traditions of the hunter-gatherers who went before them.

The zigzag or serpentine line is almost universally used as a symbol for water, and Gimbutas suggests that this was also so in both Neolithic and Palaeolithic times. Water is, of course, one of the elements of life itself without which no organism can survive. One cannot imagine any human society could remain indifferent to its vast potential for symbolic elaborations. Zigzags and meandering lines that evoke the idea of water appear to originate in very remote periods of the Stone Age, even before many people believe humans had gained the capacity to express themselves in symbolic ways. The discovery in the early 1970s of a bone fragment from the Mousterian site of Bacho Kiro in Bulgaria suggests that the use of the sign may date back to the time of the Neanderthals. This fragment of bone was engraved with the zigzag motif, and according to Alexander Marshack, who conducted a microscopic analysis of this object, it was clear that 'when the maker of the marks came to the end of an engraved line in the zigzag . . . he [or she] did not lift his [her] tool to make a joining line in the other direction, but left it on the bone and turned or twisted it, leaving the print of the turning in the corner of each of the angles'. Marshack concludes from this that the zigzag was not a random series of marks caused by cutting or abrading the bone in the course of butchery or some other practical activity, but clearly the result of the intent to create a zigzag. The zigzag is a very common motif throughout the Upper Palaeolithic period and its repeated association with depictions of fish strengthen the case for it being fundamentally a glyph for representing water. An engraved reindeer rib from the site of Cro-Magnon (from which the term Cro-Magnon man derives) at Les Eyzies in southern France, about 30,000 years old and belonging to the Aurignacian period, depicts a human-like figure with a zigzag motif on its torso.

Paul Bahn, one of the leading specialists on Palaeolithic cave art, has suggested that there appears to be a significant correspondence between decorated caves in the Pyrenees and eastern Cantabria and thermal and mineral springs in the vicinity of such sites. He suggests that this may have played a role in the mythology of these times and that the selection of such sites may not have been solely due to practical considerations. In a particular area of the French Pyrenees, cave art that includes the recurrent motif of serpentine lines is not limited to the Upper Palaeolithic period but is also found repeatedly in decoration that has been identified as belonging to Chalcolithic and Bronze Age times and may be associated with a mother-goddess cult (caves are almost universally associated in mythology with the womb). Bahn suggests that this may indicate the continuity of ritual themes through millennia of prehistory, at least in this region. Gimbutas identifies symbolic continuities from the Magdalenian period through to the Neolithic cultures of Old Europe in the depiction of zigzags and M-shaped signs in conjunction with feminine symbols, particularly the uterus and vulva. She notes that the decorated vases of Old European cultures are water containers embellished with both zigzags as symbols of water and faces representing the goddess as the source of this life-giving liquid. The zigzag also appears among the Old European signs said to be the characters of the script, and its significance can even be traced to ancient Egypt, where the hieroglyph^^ or *mu* means 'water'. Further to the east, Marshack has documented the persistence of the zigzag motif in the Stone Age art of the Russian plain, finding the depictions of the Upper Palaeolithic period echoed in the Neolithic symbolic traditions.

The single V and the chevron (an inverted V) are among the most common of the recurrent motifs in the Stone Age. The V is, of course, a very simple visual sign. It is one of the five core symbols of the Vinca sign system and also occurs among the characters of the Old European script. Clearly it is associated with the zigzag or meander symbols, which are made up of interconnected Vs and chevrons. In Gimbutas' reading, the V is often a depiction of the vulva but also had an additional connotation, of which she states: 'it is ... amazing how early this bit of "shorthand" crystallised to become for countless ages the designating mark of the Bird Goddess'.

Although we find the V used to depict birds in flight in numerous children's pictures even today – which may attest to the universality of the representation of birds by the use of the V motif – it is hard to agree with Gimbutas that the figurative images of birds (especially in relation to depictions of the goddess) and Vs concur together in the art and decorated artefacts of the Stone Age frequently enough to make this assertion. Furthermore, when Vs occur without birds and vice versa, the proposed widespread relationship is hard to maintain. Her interpretation of the V as primarily a motif relating to the vulva is also open to question. In Stone Age representations of most or all of the female body, the genitalia are usually drawn in the form of a triangle, and often with a median line. To say that isolated triangles represent the vulva is difficult enough to sustain, so Vs in isolation are even more controversial as images of vulvae. The question of the sexual nature of prehistoric symbolism is taken up in more detail in Chapter 14.

I have already described the resistance from within the academic community to an impartial investigation of the possibility that writing may date from the Neolithic. But for now we leave the Neolithic to continue the investigation by probing back further in the past to the Old Stone Age. In an attempt to revive the search for the Palaeolithic origins of writing, Allan Forbes and Thomas Crowder also highlight this institutional inertia:

> The proposition that Ice Age reindeer hunters invented writing fifteen thousand years ago or more is utterly inadmissible and unthinkable. All the data that archaeologists have amassed during the last one hundred years reinforce the assumption that Sumerians and Egyptians invented true writing during the second half of the fourth millennium. The Palaeolithic-Mesolithic-Neolithic progression to civilisation is almost as fundamental an article of contemporary scientific faith as heliocentrism. Writing is the diagnostic trait, the quintessential feature of civilisation. Writing, says I. J. Gelb, 'distinguishes civilised man from barbarian'. If Franco-Cantabrians [i.e. Ice Age inhabitants of parts of France and Spain] invented writing thousands of years before civilisation arose in the Near East, then our most cherished beliefs about the nature of society and the course of human development would be demolished.

In the early days of prehistoric research, when the dogmas of this discipline were still in their infancy and speculations concerning Palaeolithic man were rife, the suggestion that writing may date back to the time of the cave paintings was taken seriously by a number of prehistorians. Edouard Lartet (1801–73), a palaeontologist turned archaeologist, and his colleague the wealthy businessman Henry Christy (1810–65) worked together during the 1860s excavating at a number of Palaeolithic sites in southern France. Among their many achievements, they were the first archaeologists to conduct systematic investigations of sites that yielded objects displaying symbolic designs. They suggested that the carved and notched bones that they had discovered at the rock shelters of Les-Eyzies-de-Tayac in the Dordogne were records of animals that had been killed by hunters. Further investigations of these objects by Professor T. R. Jones supported their interpretation by describing the intentional markings as 'inscriptive'. In 1867 Felix Garrigou proposed that a marked artefact found at another French site (La Vache) displayed what were perhaps 'the first characters that served to represent an idea by signs'. In 1905 Edouard Piette took things a stage further, suggesting that the Upper Palaeolithic inhabitants of the region possessed not one but two forms of writing. Shortly after Piette's bold assertion, Armand Viré expressed the belief that his study of the marks on a Magdalenian artefact revealed they were also writing.

These early ideas concerning the possibility that Palaeolithic man may have invented writing were not developed into systematic studies and, as such, remained simply suggestions. The flurry of interest in this possibility waned when, with the discovery of spectacular cave paintings in the region, most prehistorians shifted their attention to other more tangible subjects. The idea of a Palaeolithic genesis for the written script never completely died out, and has been tentatively broached on a number of occasions by eminent archaeologists. Abbé Henri Breuil (1877–1961), one of the most renowned prehistorians of his generation, speculated on the possibility that abstract markings on the ceiling of the Spanish cave of Altamira were alphabetical signs. He did, however, reject this possibility after a more detailed study of the cave. The great novelist and critic Georges Bataille was also of the opinion that some of the signs found in the caves were examples of a primitive mode of writing.

André Leroi-Gourhan, whose interpretation of the cave art became the dominant one during the 1960s, did a great deal to displace previous interpretations that saw in the paintings and marks of Upper Palaeolithic man a form of hunting magic. By analysing many of the numerous groups of signs that appeared on the surfaces in both caves and rock shelters (and which, along with the naturalistic depictions of animals, make up what is called 'parietal art') and on small embellished artefacts (often grouped together under the name 'mobile art'), he was able to show that there was some system to these depictions. Leroi-Gourhan did not think that the prehistoric signs were a form of writing, but he did believe that they were a symbolic means of communication used by the people of Upper Palaeolithic times. For although he proposed that the signs represented a complex system and a considerable capacity for abstract thought, his view of their social function differs widely from interpretations that would ascribe to such signs the characteristics of writing. Peter Ucko and Andrée Rosenfeld outline the way of 'reading' of the signs according to the theory of Leroi-Gourhan:

On the basis of a detailed inventory of the 'non-naturalistic' signs . . . of

Palaeolithic parietal art Leroi-Gourhan has constructed an evolutionary series which he thinks divides up the signs into two groups which show a clear derivation either from the whole female figure and the female sexual organs, or from the male sexual organs . . . Leroi-Gourhan sees the two groups of signs as opposed, coupled or juxtaposed; those of [group] 'a' (lines, dots, etc.) representing the male, and those of [group] 'b' (ovals, triangles, etc.) representing the female.

In this reading of the signs, with its division of the male and female, we can see certain similarities with Marija Gimbutas' 'alphabet of the metaphysical', and in fact, Leroi-Gourhan even described the symbols as a 'metaphysical system'. But the similarities are superficial. For whilst Gimbutas' reading of the Neolithic symbols and their Palaeolithic precursors presents an essentially feminine alphabet with a highly charged vibrant emotional and spiritual backdrop, Leroi-Gourhan's intellectual approach presents a highly structured and cold system in which the sexes are balanced in a harmony which is more mathematical than musical, more cerebral than spiritual. But the indications are that towards the end of his life he began to be convinced that the Upper Palaeolithic symbols were more advanced than he had previously imagined. According to Brigitte and Gilles Delluc: 'with reference to the complexity of the signs at Lascaux, Leroi-Gourhan admitted to us shortly before his death, "At Lascaux I really believed they had come very close to an alphabet".' The importance of this should not be underestimated. If Leroi-Gourhan, who was one of the most respected prehistorians of his generation, believed they had come so close to developing a system of writing, then clearly the investigation of the possibility that they went further than that and actually invented an alphabet is worth considering and ceases to seem such a wild and impossible notion. It also strengthens the case for looking hard at Neolithic evidence such as that from Old Europe in order to detect possible writing systems. If Leroi-Gourhan was right that Ice Age cultures came to the very brink of developing an alphabet, it would be rather surprising if no further developments took place until the rise of the Sumerian script.

Forbes and Crowder's justification for reviving the idea that writing may perhaps be traced back to the Ice Age is based on the fact that a considerable number of the deliberate marks found on both parietal and mobile art from the Franco-Cantabrian region are remarkably similar to numerous characters in ancient written languages extending from the Mediterranean to China. *Figure 16* shows a selection of Franco-Cantabrian signs and *Figure 17* gives examples of characters from the following: (a) hieroglyphic determinatives; (b) Sumerian pictorial writing; (c) Indus Valley; (d) Linear A; (e) Linear B; (f) Cypriote; (g) Proto-Sinaitic; (h) Phoenician; (i) Iberian; (j) Etruscan; (k) Greek (western branch); (l) Roman; and (m) Runic. *Figure 18* shows a number of signs found on the oracle bones, the earliest widely accepted instances of the Chinese system of writing. Whilst the comparatively high

Figures 16, 17, 18

degree of correspondence between the Mediterranean signs can be easily explained by pointing out that some of the written languages drew direct inspiration from others (which is undoubtedly the case), the far-flung similarities across time and space – from the Ice Age to historical times and from Spain to China – defy such a simple and straightforward explanation.

It is not as easy as it might at first appear to dismiss these apparent correspondences as simply due to chance, to coincidence. It can be argued that there are only so many basic signs that humans can come up with and that this explains the 'mirage' of any cultural links said to exist between sign groups from diverse times and places. Even if this were the case, then the fact that there is some definable limit to the human inventory of 'abstract' signs irrespective of culture would be an avenue well worth pursuing. It may imply that there is some cognitive mechanism which lies behind the generation of the visual form of such signs, behind the diverse meanings which the signs impart in various cultural settings. We cannot, however, in the present state of our knowledge of the past (particularly of prehistory), categorically rule out the possibility that certain similarities between sign inventories that are currently believed to have no cultural connections may in fact be connected after all. We have already seen how no less a figure than Sir Flinders Petrie had pursued some aspects of the unexplained links between ancient Mediterranean signs.

Forbes and Crowder have suggested that the Franco-Cantabrian marks may be remote ancestors of the prehistoric pottery marks from Egypt and also of the early Mediterranean written languages. Given the prevalent current view that Upper Palaeolithic humans were too 'primitive' to be able to invent writing or, at best, simply had no 'need' for writing, the serious investigation of such an early possible origin for the art of writing has not been carried out. Again we can see the economic argument for the origin of Sumerian rearing its head. There are no firm grounds for asserting that writing necessarily emerges as a consequence of economic requirements. As has already been said, the Chinese oracle bones owe their emergence to an interest in pursuing divinatory practices and not because of the necessities of accounting procedures. The Chinese had, according to the restrictive model of economic necessity, no need to invent writing for themselves. Nevertheless, they did develop it and so there is no need to think that other cultures – including those of Ice Age Europe – did not develop a system of writing for divinatory or other 'magical' purposes. I do not mean that Ice Age humans must have invented writing; I am simply saying that there are no acceptable a priori reasons to reject the possibility out of hand. Neither their hunting-and-gathering lifestyle nor their belonging to remote prehistory should deter us from seeking to press on in our attempts to understand when, how and why writing began.

Paul Bouissac sees the resistance to the serious investigation of the possibility of Palaeolithic writing as partly due to an entrenched tradition of viewing Ice Age paintings and other forms of prehistoric art as simple representations of the objects that they depict. In other words: 'the painting of a bison, for instance, unambiguously indicates that its crafter(s) meant to refer to a bison as such. This can be called the representational fallacy.' In such a representational way of seeing things, the abstract signs of Palaeolithic art are seen not as abstract at all but rather as representations of hunting nets, weapons, etc. Whilst it is very likely that some of the art of the Upper Palaeolithic period was indeed purely representational, we need not presume it was all so. He suggests that a more fruitful approach is one he calls the semiotic hypothesis, which he describes as follows:

> The semiotic hypothesis can be subdivided into two formulations as is often done in the social sciences: the weaker hypothesis (individual visual designs or their relations have meanings which are to an extent determined by at least some of the qualities of the animals or artefacts represented) and the stronger hypothesis (such resemblances are accidental, historical, or purely instrumental – i.e. no more relevant to the content of the message than the shape of the wedge in the syllabic cuneiforms of Mesopotamia is relevant to the meaning of the messages this system of writing can convey).

In this way of looking at things, the 'alphabet of the metaphysical' of Marija Gimbutas (which she saw as preceding the Old European script) and the system of sexual symbolism developed by Leroi-Gourhan would be examples of the weaker hypothesis, since both their interpretations of Stone Age signs emphasise the representational element of various motifs as being an important factor in understanding the message that prehistoric people were trying to convey. The stronger hypothesis, which basically seeks to test sets of signs to see if they are comparable to what is generally understood and accepted to be a system of writing, is the approach advocated by Bouissac. In promoting such an approach, he does not presume that writing systems will be uncovered; he only recommends that the hypothesis be tested. There are many practical obstacles in the way of conducting such research, as it:

> presupposes the availability of a database through which the morphological characteristics of large sets of signs in their original configuration can be probed in order first to identify semiotic boundaries across both space and time; second, to construct types from the systematic study of morphological variations; and third, to ... scan strings of signs in whatever directions sequences can be construed, in search of recurring combinations of types. Then, consistent formal properties of such type combinations may or may not emerge. In the latter case, the demonstration of pure randomness would have been made, and, consequently, would allow us to discard, at least for the time being, the strong semiotic

hypothesis. But if a sufficient degree of formal consistency were revealed, a new direction of research would then be open.

Bouissac is making the point that if a large enough corpus of Upper Palaeolithic signs can be transcribed in a way that retains their original spatial layout and shows them to belong to a particular period of time (rather than be an amalgam or jumble of signs painted or inscribed at separate times), the investigation into whether the corpus is a form of writing or not can proceed. There are many sequences of signs preserved at the Franco-Cantabrian sites, but whether they are sufficient in number to permit an analysis such as that proposed by Bouissac is unclear. Whether such a hypothesis is ever put to the test in seeking to understand the sign systems of the Upper Palaeolithic age remains to be seen. At present most scholars seem happy to pursue a more cautious approach by interpreting the various signs as a partly abstract symbolic means of communication but not as a fully formed system of writing.

In the last few chapters I have selected only a small number of the complex sign systems that have been preserved from prehistoric times. My concentration on the Near East and more particularly on Europe should not be taken to imply that such systems did not exist elsewhere in the prehistoric world. Far from it; investigations of numerous collections of signs are being undertaken in places as far afield as the Arabian peninsula, China and Australia. Millions of prehistoric signs across the continents have already been recorded, and more and more are being discovered all the time. Most researchers would agree that these countless signs are, in the main, not just the mindless doodlings of prehistoric man, but the problem of interpreting them remains an elusive and frustrating enterprise. It no longer seems sufficient to retain a simplistic evolutionary sequence of events leading up to the Sumerian breakthrough some 5,000 years ago.

In the course of my discussion of the theories of Schmandt-Besserat concerning the Neolithic influence on the emergence of the Sumerian script, I mentioned her ideas on the Palaeolithic precursors that provided the initial evolutionary steps upon which the Neolithic token system was ultimately built. In her view, the tallies of the Upper Palaeolithic period were more rudimentary means of storing and manipulating information. Thus, whilst her theory has challenged the clear-cut divide between 'civilised' cultures that can boast writing systems and 'primitive' prehistoric cultures that were entirely ignorant in this respect, it maintains a rather traditional view of a simple evolutionary development in complexity from Palaeolithic to Neolithic. Concerning the immediate ancestry of the Sumerian script, we can only agree with the endorsement of Schmandt-Besserat's theory by William W. Hallo, William M. Laffan Professor of Assyriology and Babylonian Literature, and Curator of the Babylonian Collection at Yale University:

Whatever challenges have been or are yet to be encountered by the thesis, however, these would have to offer an equally systematic alternative to be convincing. The sudden appearance of the sophisticated script of the archaic texts from Uruk *ex nihilo* and *de novo* is an argument sustained neither by reason nor by the evidence. *Before Writing* furnishes to date the most coherent working hypothesis to account for the prehistory of the historic invention known as writing.

Yet the shortcomings of previous approaches to the Neolithic evidence (namely that Neolithic people were not 'advanced' enough to develop any complex devices for encoding and dealing with information) that she has almost single-handedly overcome with her discovery of the token system are also to be found in her own treatment of the earlier evidence. Her approach to the available data concerning the capacities of Palaeolithic peoples is based on a fallacy of evolutionary thinking, namely that earlier always means simpler. Not all Upper Palaeolithic symbolic systems are as simple as the tallies she portrays as typical of this period. In a highly sophisticated and rigorous analysis of an engraved reindeer antler belonging to the Magdalenian period (estimated to be more than 14,000 years old) discovered at La Marche cave at Lussac-les-Châteaux in western France, Francesco d'Errico has demonstrated that more complex systems certainly existed during the Upper Palaeolithic (see *Figure 19*). He explicitly refutes the notion that artefacts from this time could only store information of one type:

> The marks on the La Marche antler seem to indicate that artificial memory systems which could record different categories of information existed well before the systems of tokens evoked by Schmandt-Besserat. The use of complex artificial memory system codes in the Upper Palaeolithic is confirmed by the study of an engraved pendant found in the final Upper Palaeolithic layers of the Tossal de la Roca shelter near Alicante in Spain. On this object, as on the La Marche piece, we observe changes of point and technique between rows of tiny incisions arranged in neighbouring rows ... Thus, the La Marche antler indicates that the problem of the origin of writing cannot be correctly addressed from an archaeological point of view without taking into consideration the evolution and variability of Palaeolithic artificial memory systems. It is likely that the course of this evolution was more complex than we have previously imagined.

Furthermore he points out that in some cognitive respects the La Marche antler is actually more complex than the later, and thus supposedly superior, Neolithic token system. D'Errico suggests that cognitive evolution must be reconsidered in the light of his study of the La Marche antler and analogous artefacts from the Ice Age, and he also raises the question of what we are

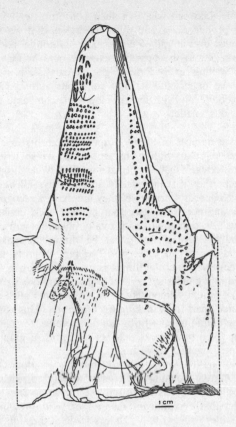

Figure 19

actually discussing when we talk about writing and its origins. He takes issue with traditional ways of defining writing which would automatically assign objects such as the La Marche antler to the pre-writing phase in an evolutionary framework. A fairly representative example of conservative thinking of the evolutionary position is found in *Writing* by David Diringer, a book on the development of writing:

Without writing, culture, which has been defined as 'a communicable intelligence', would not exist (except, perhaps, in a form so rudimentary as to be virtually unrecognisable) . . . We can perhaps say that all forms of graphic inscription, however crude or refined, have their roots in the central and universal human need to *communicate* and *express*. Nevertheless, a clear distinction must be made, if our subject is to be rendered at all practicable, between what we shall henceforth call *embryo-writing* and *writing proper* ... a line must be drawn, and we should not, in reacting against oversimplified ideas about 'primitive' and 'rude' stages of civilisation, go to the opposite extreme, and lump together in the category of 'writing' every form of graphic expression used by man ... Writing, as we understand it, is a conscious activity, intricately and inseparably bound up with the development, comparatively recent, of man's conscious intellect.

Diringer makes a number of highly suspect statements in even this short passage. He states that culture barely exists in societies – whether prehistoric or historical – that do not have writing. *No* anthropologist would accept this position, as it is universally accepted that all human communities, however supposedly 'primitive' they are, have a recognisable cultural inheritance. Furthermore, the idea that humans lack conscious intellect unless they have a system of writing is totally fallacious and absurd. Finally, Diringer's comment that a line must be drawn to make his subject practicable is also open to criticism. The fact that his task – namely describing the development of writing – is made easier by dismissing Upper Palaeolithic and many other systems of signs as embryo-writing cannot be denied, but what is really important is to ask whether such a division has any genuine value apart from making his life easier. This is not to say that his views on the definition of writing are of no value, for he touches on an important point when he remarks that not all forms of graphic expression should be lumped together as writing.

There are two possible ways forward from this position. One alternative is to extend the current meaning of the term 'writing' to at least some forms of what have been dismissed as 'pre-writing' or 'embryonic writing' in the light of the fact that evolutionary theories of writing cannot adequately explain what divides 'true writing'. Unless some of the sign systems of the Upper Palaeolithic or the Old European script are shown to be kinds of writing that would fit into conventional understandings of what is 'true writing' (if either were shown to be so, then the entire evolutionary sequence from prehistoric 'pre-writing' to civilised 'true writing' would, of course, collapse overnight), then it may be better to seek a different way to investigate prehistoric sign systems that can be liberated from the ideological straitjacket that always relegates such symbolic behaviour to the most lowly rungs of the evolutionary ladder. This would be the second alternative and one that is advocated by d'Errico. He suggests that the uniquely human capacity to develop means by which information can be recorded, manipulated and communicated outside the confines of the body manifests itself in both prehistory and history in numerous forms which can be called artificial memory systems (AMSs). Upper Palaeolithic tally sticks, the La Marche antler, the Neolithic token system, Sumerian writing, the Tartaria tablets and even computers are all examples of AMSs. By making use of this term we are able to avoid simplistic and misleading evolutionary doctrines which get in the way of assessing what kinds of symbolic codes existed in prehistory and what their functions may have been.

Marshack, Gimbutas and many others have made it clear by intensive and extensive studies that Stone Age symbolism cannot be reduced to the idle scribbling of simple shapes that have no intrinsic meaning. The recurrence of essentially similar signs over the whole of Europe (and, of course, beyond) and throughout the Upper Palaeolithic, Mesolithic and Neolithic periods represents a still barely understood set of systems of symbolic communication. As these signs occur throughout the continent and over a period of at least 25,000 to 30,000 years, it would be foolish to suppose they were all part of a single 'metalanguage' that was commonly used and understood across this vast time and space. This is not to say that Gimbutas' interpretation of some of them as symbols of a goddess-worshipping religion, nor Marshack's explanation of numerous artefacts as lunar calendars (a theory discussed in the next chapter), is incorrect. Both these researchers have brought to light aspects of the great corpus of Stone Age signs, but the vast project of decipherment is still in its infancy. So in the preceding few chapters we have followed our quest for the origins of writing all the way from Sumeria to Ice Age Europe. It is now more clear that the writing systems that emerged in Mesopotamia and elsewhere around 5,000 years ago owe their development to innovations that can be traced back to the Neolithic period and, in various respects, even as far as Upper Palaeolithic times.

Chapter 6

Palaeoscience

One of the fundamental and recurring themes of this book is the idea that in seeking to understand the human story it is not possible to do so with any degree of accuracy without taking into account the innovations and developments that occurred in the Stone Age. Writing and other artificial memory systems have been shown to have their roots in the Old Stone Age, and I shall now turn my attention to the primordial origins of mathematics and the roots of science. The history of science is necessarily foreshadowed by developments in the Stone Age, and as such it is necessary to look to prehistory in order to assess what occurred in this area of human endeavour prior to the advent of what is commonly understood to be civilisation. Since writing has been shown to be the outgrowth of a long tradition of notation and symbolic sign systems, it would be foolish to think that scientific awareness appeared overnight. Therefore the history of science needs to be augmented by another area of research that may be called the prehistory of science. Despite the obvious requirement to account for the development of scientific thinking and practice, it is generally assumed that little or nothing happened in the most archaic periods, and many historians of science seem to be under the impression that, like civilisation itself, it all began 5,000 years ago and before that only the most rudimentary and insignificant activities took place.

There are few exceptions to this rule by which science is seen as exclusively an attribute of the 'true' civilisations. The most notable exception is in astronomical knowledge which is widely thought to have been part of the intellectual culture of Neolithic peoples. This has become widely known due to the considerable number of popular books on the astronomical alignments or orientations of megalithic monuments. Astronomers have long had an interest in Stonehenge and other sites. The astrophysicist Sir Norman Lockyer made accurate observations of Stonehenge in the first few years of the twentieth century and concluded on the basis of these that it was a solar temple in which astronomical observations that were made by its architects were subordinated to ritual considerations. The interest in Neolithic astronomy reached a zenith in the 1960s, largely due to the writings of the engineer Alexander Thom. Thom had been conducting research along these lines since the 1930s, and the results of surveys he had undertaken at a considerable number of sites were initially announced in papers published in the 1950s and in a more comprehensive and accessible form in his first book on the subject, *Megalithic Sites in*

Britain, in 1967. In this carefully researched book Thom claimed that the makers of the megaliths made detailed observations of the movements of the sun and the moon and had a considerable degree of skill as engineers and in the practical applications of geometry. Although reactions to the book were somewhat mixed, it was recognised even by sceptics as a serious contribution to the study of the origins of science. A whole host of publications on the subject subsequently appeared and arguments over the extent, accuracy and purpose of the alignments of Neolithic (and Early Bronze Age) sites continue to this day.

There is no doubt that astronomical observations were made in the Neolithic period, but whether they constituted science as we understand it today seems unlikely. This in no way denigrates the striving by Stone Age people to observe, as accurately as they were able to do, the movement of the sun and moon. One cannot directly compare their approach to the study of the heavens to that of modern astronomers, for such Neolithic observations were intimately bound up with ritual activities and were thus integral to their cultural lives in ways which must seem mysterious to the secular scientist. As the Neolithic evidence has become widely disseminated, I shall not dwell on it here. Instead I shall go even further back, to a time scale in which these Neolithic contributions to scientific knowledge seem distinctly modern, back to the primordial functioning of the minds of the Palaeolithic pioneers of knowledge.

Prehistorians have long understood that even hominids were able to transmit knowledge through the generations, and in so doing they created their own traditions, of which we can glean a little from the evidence of the vast number of stone tools of Lower Palaeolithic age which have survived hundreds of thousands of years. The hand-axe is the most characteristic tool of the Acheulian technological tradition, which existed for more than a million years. Many archaeologists have commented upon the fact that hand-axes are highly standardised in their dimensions and form across the continents of Europe, Africa and also in many parts of Asia. This is not, of course, to say that they are all exactly the same but that, allowing for various factors such as the idiosyncrasies of the individual craftsman who made them, the various different raw materials out of which they were fashioned and the existence of regional styles, they do indicate a remarkable degree of uniformity. Clearly this repetition of the form and dimensions of the Acheulian hand-axe is not due to chance but rather to the communication of knowledge. This body of knowledge was larger and more long-lived than any individual tool-maker and so it constituted a social form of knowledge. V. Gordon Childe described the standardised tool as 'a fossil idea' that required a certain capacity of abstract thought on the part of its makers. That is to say that in order to reproduce the standardised tool again and again, its makers must have had some sort of an image of 'a tool in general' rather than merely perceiving each concrete instance of a tool separately. It seems that not only did the Acheulians have some capacity for abstract thought, but the origins of aspects of mathematical knowledge may be traced back to the time of the hand-axes. The Palaeolithic archaeologist John Gowlett has this to say on the matter:

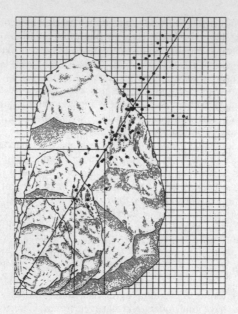

Figure 20

In this diagram the diagonal line represents a perfect correlation between length and breadth of a hand-axe or cleaver. The dots furthest away from the line denote the most irregular specimens. All these hand-axes came from one area of the Kilombe site and the great majority were made in proportion so that small and large specimens are the same shape.

Hand-axes from many ... sites, show that early man could employ the same 'template' in making tools of different sizes. In other words, the mental apparatus already existed for making basic mathematical transformations without the benefit of pen, paper or ruler. Making hand-axes in the same proportions at different sizes provides the earliest practical demonstration of principles treated hundreds of thousands of years later in Euclid's *Elements of Geometry*. Similar abilities are fundamental to the practice of visual art. Could such templates be passed from person to person, from mind to mind, by visual copying alone? All our present experience of using such skills suggests that language would be required.

To illustrate his point he made a detailed examination of hand-axes from a particular area of the Kilombe site in Kenya which date to about 700,000 years ago. The size and shape of the various hand-axes that made up the sample were plotted on a diagram (*Figure 20*) and as the explanatory note makes clear, most of these tools conformed to a proportional norm. One researcher who has undertaken detailed study into the prehistoric roots of mathematics is the Russian prehistorian Boris Frolov, who has sought to track down the origins of numerical operations in the Stone Age even as far back as the Lower Palaeolithic period. Like Gowlett he has seen the standardisation of the hand-axe as evidence of at least a level of abstract thought, and notes that the symmetry of the hand-axe is a significant feature absent in the cruder pebble tools that represent the earliest kinds of tools that preceded the Acheulian period. Frolov does note that the tool-makers of the pre-Acheulian period must have been aware of the symmetrical forms that existed in nature and, of course, in their own bodies, but also notes that they did strive to achieve such an effect in their tool-making. This quest for symmetry which occurred in the Acheulian period was accompanied by a more pronounced and developed rhythmical sense applied to technological operations. He also sees the technological refinements that took place in the Lower Palaeolithic period as accompanied by a more conscious understanding of the operations of division, made particularly manifest to early man in the sphere of tool-making, which basically involved the splitting and flaking of a whole stone into a number of parts consisting of the produced tool and the associated waste flakes that were produced during the manufacturing process.

The next stage in the embryonic development of mathematics is seen by Frolov to have occurred in the Middle Palaeolithic (with its characteristic stone tool industry known as the Mousterian) and is signalled by the making of groups of parallel lines on various bone artefacts dating from this period (the whole question of the Lower and Middle Palaeolithic evidence for non-utilitarian and symbolic activities is dealt with in Chapters 16 to 18). An important step forward has been seen by Frolov in this ability to cut straight lines and arrange them in series, for not only is it evidence of greater abilities of measurement than is known from the previous phase of prehistory, it also announces in concrete form the abstraction and externalisation of a particular element – namely the straight line. Whilst the Acheulian hand-axe of the Lower Palaeolithic indicates that its makers must have had a notion of a tool in general, they did not depict this abstract notion in any concrete form. In contrast the Mousterian maker of straight lines was able to separate his or her understanding of the concept of straightness from its particular occurrence in a concrete form. As Frolov put it, the Mousterian carver:

> used more refined and complicated comparative measurements than any known prior to that period; the unit of such comparisons was, for the first time, an element isolated by him – a straight line, abstract in the sense that it was isolated for the sake of that geometrical property from all other properties of objects created up to that time (for example, the straight blade of a tool, a straight haft, etc.) . . . the object of these comparative

measurements was a series of identical and specially created components differing in only (a) their position along a given portion of the surface of the bone, (b) their position relative to the nearest similar elements, and (c) the number of elements in the group of which it is a component.

In the development of the practice of burial, which occurred first in Middle Palaeolithic times, Frolov sees the first inklings of astronomical observations. There are some indications that orientation of the corpse was practised among Neanderthals, and if this was indeed the case then it would indicate an understanding of the four cardinal directions and the motion of the sun. Clearly if these observations could be confirmed, then they would also indicate an awareness of time as much as space, as the alternation of night and day and the dawn and the twilight would necessarily have been part of early man's observations at this time. A knowledge of the seasons would have been of considerable importance in the Middle Palaeolithic period, as the migration of game and the preparation for the more hostile weather of the winter season would have been matters that the people of this time could scarcely afford to ignore. But how much of such knowledge was instinctive, as it is in animals, and how much it can be said to reflect a nascent understanding of quantity and a uniquely human understanding of the world is hard to discern based on the available evidence. Frolov's whole reconstruction of the Lower and Middle Palaeolithic roots of mathematical thinking is necessarily speculative, but the tool-making capacities of early man far exceed the abilities of other animals, and the levels of precision, symmetry and forethought that went into the manufacturing of tools indicate a considerable degree of skill and abstract thinking.

With the beginning of the Upper Palaeolithic period, in which the emergence of behaviourally modern humans is generally thought to have occurred, the reconstruction of the origins of mathematics becomes somewhat easier, as the archaeological record becomes much richer. This is partly due to the fact that geological and other natural forces have obliterated – almost entirely – any objects made in the Lower and Middle Palaeolithic periods other than those most hardy and robust of artefacts, namely stone tools. The art of the Upper Palaeolithic gives us an insight into the use of scales and proportions during this time. In both paintings and figurines from the period one can see that representations of both humans and animals reproduced their subjects in a great variety of scales. Perhaps the most extreme example of the small scales often used in Upper Palaeolithic art comes from the discovery of a figurine depicting a mammoth which was made of a kind of stone called marl. This minute figurine, found at the site of Kostenki XI in Russia, is only $10 \times 8 \times 5$ millimetres in size, the scale of which has been calculated to be a reduction of 1:400 of the real-life mammoth. Frolov has noted that in the upper Palaeolithic period there is also a use of dimensional perspective and 'other kinds of perspective described by Leonardo da Vinci'. But the most important development that he sees as identifiable in the Upper Palaeolithic evidence that is not apparent in the earlier periods is the concept of number.

In order to follow the next stage of Frolov's argument it is necessary to digress a little and investigate the practice of counting among tribal peoples, from which he infers how people in the Upper Palaeolithic may have counted and also what kind of artificial memory systems they used in order to provide a more permanent record of such procedures. Many people still believe that tribal peoples were unable to understand numbers beyond the rudimentary level of having concepts of 'one', 'two', and 'many'. Whilst some languages spoken by tribal peoples do not have words for numbers beyond two or three, this does not mean they were unable to count any higher than that. Some cultures simply had little need to count beyond the first few numbers and this does not reflect any lack of intellectual capacity to do so. The fingers and other parts of the body were used to calculate comparatively large quantities; for example, the Chukchi reindeer herders of north-eastern Siberia (who had a need to count their herds) used a quinary-vigesimal system, i.e. one built on units of five and twenty. The fingers of one hand made up the unit of five and the sum of all a person's digits the unit of twenty. Five persons represented the number 100, and they were reported to count up to 1,000 in such a fashion.

A very similar method was used among the Yukaghir, who were also reindeer herders, and a herd of 94 reindeer was expressed as 'three persons, on top of that one person, and also half a person and also a forehead, two eyes and a nose'. Other native peoples of north-eastern Asia had a considerable repertoire of numerical terms. In the languages of the Kamchadal and Koryak peoples there were terms for numbers as high as 100; among the Ainu people of northern Japan and southern Sakhalin numbers went up to 1,000. The Aleuts of the Aleutian islands off the coast of Alaska were reported by an early Russian visitor to be able to count to over 10,000, and were even reported to be able to denote quantities in the millions! Whilst the Aleut case may be said to be a particularly remarkable one, the ability to count very large quantities was by no means restricted to the peoples of the far north but has also been recorded from other parts of the tribal world.

Despite the fact that the Pomo Indians of California are not usually included among the few native American cultures widely considered to 'deserve' to be called civilisations, those who have studied their culture in detail have thought otherwise. The anthropologist Edwin M. Loeb stated that 'The Pomo skill in counting was merely one of the many intellectual developments of the complex Pomo civilisation.' They seem to have developed their vigesimal system of enumeration in order to be able to count the great numbers of beads that were collected together at important ceremonial occasions such as funerals and peace treaties. By simply using sticks they were able to count many thousands of beads; Loeb's informant said that he had personally witnessed counts of at least 20,000 beads, and the Pomo word *xai-di-lema-xai*, known as 'the big twenty', refers to the number 40,000. These few examples clearly establish that the ability to count and record large quantities was not uncommon at all in the tribal world.

In Chapter 3 we have seen that there is clear evidence that a system of accounting existed throughout the Neolithic period in the Near East. It now remains to consider the question of whether such arithmetical procedures existed in the Upper Palaeolithic period. Clearly it is impossible to prove or disprove whether counting was undertaken at this time unless we can find concrete evidence of the use of artificial memory systems of some kind. Tallies and other AMSs are well attested in tribal cultures in various parts of the world, and in the last chapter I briefly alluded to the existence of such permanent records in the Upper Palaeolithic period, something I shall now return to in more detail. There have been many artefacts from this period of the Stone Age that have been suggested to be some kind of AMS, but for the sake of brevity I will mention only three particular finds – two from Belgium and the other from England – which belong to the tail end of the Upper Palaeolithic period.

The two Belgian artefacts that have been interpreted as AMSs were both found at the same site, La Grotte de Remouchamps in the province of Liège, but by two separate excavators, the first by the archaeologist E. Rahir in 1902 and the other in 1970 by Michel Dewez. Radiocarbon dating of the site puts their age at over 10,000 years. Both are fragments of animal bone that have been engraved with markings that Dewez has identified as evidence for the existence of a numerical system. The markings on the first bone are shown in *Plate IX* (the scale is in centimetres). Seven groups of five holes (in a pattern identical to that used on modern dice) are clearly visible, and on the left are a number of other groupings that have unfortunately survived only in fragmentary form. Dewez has described those on the left as (starting from the bottom) six groups of five marks followed by one group of eight, with the two upper groups being too fragmentary to be counted. As can be seen in the right-hand photograph of this object there are also a number of notches along the edge of the bone. The second bone (see also *Plate IX*) is just over seven centimetres in length and is a narrow polished splinter that has been engraved with lines on both sides (the photograph to the left represents the front of the object and the one to the right the reverse; it must be borne in mind that this designation of front and back is somewhat arbitrary, as it is impossible to say how this aspect of the object was perceived by its maker, and so this designation is simply used in order to discuss the artefact). The front has been engraved with four sets containing different numbers of marks. These groups are clearly shown in the diagram *Figure 21* and can be enumerated as follows:

A1: 14
A2: 8
A3: 11
A4: 7

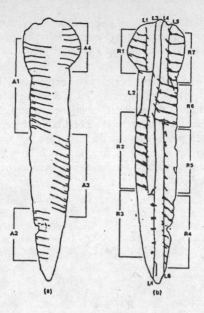

Figure 21

The back of the bone has a more complex arrangement of marks, and these are rather difficult to discern from the photograph alone but can more easily be seen in *Figure 21*. As can be seen from this, Dewez has designated the seven sets of horizontal markings and the six sets of vertical markings. The number of marks in each of these groups is as follows for the seven sets of horizontal markings:

R1: 5 + 1
R2: 8
R3: 6
R4: 8
R5: 6
R6: 5 ̄
R7: 6

And by calculating the number of horizontal lines that join the vertical lines we get:

L1: 5
L2: 0
L3: 9

93

> L4: 11
> L5: 6
> L6: 14

Dewez suspected that both the bones were expressions of some kind of numerical system, perhaps of the same kind. Starting with the first bone, which clearly emphasises the especial importance of groups of five, he used this fact in his attempt to interpret the second bone object. Dewez pursued this in two separate procedures. In the first case he simply divided the numbers in each group into sets of five, to which he added the units left over. This arrangement leads to the following breakdown of the markings on the front of the bone:

> A1: 5 + 5 + 4
> A2: 5 + 3
> A3: 5 + 5 + 1
> A4: 5 + 2

And on the back:

> R1: 5 + 1
> R2: 5 + 3
> R3: 5 + 1
> R4: 5 + 3
> R5: 5 + 1
> R6: 5
> R7: 5 + 1

The second possible reading involves applying the quinary grouping on the back of the bone not to the horizontal sets but to the meeting of the horizontal and vertical lines. This produces the following:

> L1: 5
> L2: -
> L3: 5 + 4
> L4: 5 + 5 + 1
> L5: 5 + 1
> L6: 5 + 5 + 4

Having completed this rearrangement of the markings based on the hypothesis that they are examples of the Palaeolithic use of the quinary system, Dewez notes that the front of the second object can be seen to display two relevant patterns. The first is the occurrence of 1, 2, 3 and 4 once each in the four groups of marks broken down by the application of the quinary system. Another possibly significant order is the two 'double five' numbers and two 'single five' numbers into which the four overall groups can be divided; this can be expressed:

Double	Single
5	
+ 4 (A1)	5 + 3 (A2)
5	
5	
+ 1 (A2)	5 + 2 (A4)
5	

In this arrangement the total of the 'double five' numbers would be 25 and that of the 'single five' numbers 15, both of course being multiples of five.

Concerning the markings on the back of the object in the light of the second of the two possible interpretations given above, the same ordering of the sets into 'double' and 'single' fives yields the following:

Double	Single	Five exactly
5		
+ 4 (L6)	5 + 1 (L5)	5 (L1)
5		
5		
+ 1 (L4)	5 + 4 (L3)	
5		

The total of the 'double five' numbers would be, as on the front of the object, 25; and similarly the total of 'single five' numbers would also be 15. The missing 'group' (L2) has been proposed as possible evidence for the existence of zero at this time. It seems that Dewez's analysis cannot be reduced to a simple juggling with the figures, but rather the evidence for the use of the quinary numerical system in the Upper Palaeolithic period seems to be strong. Its existence at this time should not really surprise us, as there is no evidence to suggest that the peoples of the Upper Palaeolithic period were in any way the intellectual inferiors of modern humans, and as we have seen, the counting systems used in tribal cultures down to recent times may also be rather complex. For our present purposes the actual existence of a numerical system that seems to be demonstrated by analysis of these bones is the most important aspect of these artefacts. But they must also have had a more specific purpose; that is to say they must have been used to count something, but what this something was we can only guess.

Dewez puts forward the plausible suggestion that they may have been used in playing a game of chance, arguing, quite correctly, that games of various kinds would have been a feature of life in the Upper Palaeolithic as they have been throughout the world and throughout history. He sees other artefacts found at the cave to be best explained as other pieces of gaming equipment. A group of 45 fossil shells from the site may be counters or gambling chips, and another bone from the site has been interpreted as a four-sided dice (our own six-sided dice is by no means the only kind; the four-sided variety was used by numerous peoples including the Romans and various native American peoples). Concerning the latter, Dewez has suggested that the quinary system could also have been applied to this object if the numbers 1,2,3 and 4 were arranged on its four sides with the 1 opposite the 4 and the 2 opposite the 3.

Similar evidence for the existence of a quinary numerical system comes from the English discovery that is estimated to be around 12,000 years old and was found in the late 1920s at Gough's Cave in Somerset. A number of tools made out of the leg-bones of hares were found at the site, one of which had subsequently been marked with a series of notches. Although this particular artefact had a polished appearance, this occurred prior to its use as an AMS and must either have been part of the manufacturing process or a result of wear in the time it was used as a tool. This conclusion was reached by E.K. Tratman, who made a close study of this artefact in the 1970s, on the grounds that no sign of any use wear was evident on the notches themselves, unlike the rest of the surface of the object. The significance of this will become clear below. The natural form of this hare bone means that it has three borders (or ridges) along its length, and the prehistoric individual who marked the bone did so by making the notches at right angles to these three border lines. Tratman refers to these three borders as anterior, internal and external. *Figure 22* shows Tratman's schematic reconstruction of the grouping of the notches on the bone. The internal border clearly exhibits – starting from the 'head' of the bone – groups of 5-4-4-4 and then, after a damaged area, another group of 4. He has indicated on the diagram that the broken section on the internal border probably consisted of three groups of four. The notches on the anterior border are intact and are in a grouped sequence of 6-4-4-4-4-6. The external border is, like the internal, damaged, and whilst five groups of five notches are clearly visible, the number of notches at the head end is unclear. He has suggested that the one visible notch nearest to the head end was most likely to have been accompanied by four others, thus making a total of six groups of five notches (although he accepts that other reconstructions of this particular group of notches are possible).

Concerning what the purpose of the notches might be, Tratman rejects the idea that it could have been a gaming piece of the kind suggested for the Belgian artefact by Dewez because no use wear is evident on the Gough's Cave bone. It is possible that the latter could have been an item used for recording the game – in effect a sort of prehistoric score card – as this would not entail its regular handling and usage that would result in use wear. This

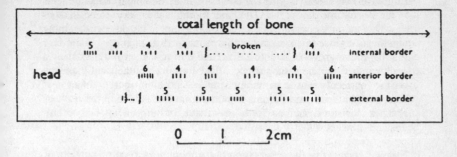

Figure 22

is of course only a suggestion, and there are numerous other possibilities that would be equally plausible explanations for the notched bone, but Tratman believes that whatever its specific function, it was essentially an instance of the use of some system of numeration, and he even refers to it as a 'calculator', concluding 'that some relatively complex form of numeration is exhibited. If [this is] so then the late Upper Palaeolithic people had a much higher capability in the use of numbers than is generally attributed to them and it follows that there were concomitant terms within their language(s) to explain the system(s) in use.'

Time represents one of the most difficult things to count without the aid of an AMS of some kind, and the existence of calendars has also been proposed for the Upper Palaeolithic period. People in the Old Stone Age were obviously aware of the motion of the sun and the moon, and it is the latter of these heavenly bodies that is the most likely to have been the main subject of Palaeolithic time reckoning, because it involves a smaller numerical sequence than that required to record the solar cycle or year. It is the observation of the lunar cycle by Upper Palaeolithic peoples that Frolov believes to be the key to the understanding of the nature of archaic cosmologies:

The moon is the largest heavenly body that an inhabitant of the earth can observe with the naked eye. Moreover, no other celestial luminary undergoes such marked changes in its visible shape as the moon. In the lunar cycle it is possible to identify several obvious limits, demarcations that may have caught the attention of ancient man earlier than others. Every 27 days, 7 hours, and 43 minutes the moon returns to its former visual position among the stars (the sidereal month). Recurrence of the full moon, the new moon, and other phases occurs at intervals of 29 days, 12 hours, 44 minutes (the synodical month). After the new moon the first quarter-moon appears after 7 days, 10 hours, the second after 14 days, 18 hours (full moon), the third after 22 days, 3 hours, and the fourth is the following new moon. During the new moon this luminary cannot be seen in the sky for one day or two. The lunar month is thus taken to be 28 days ... The moon draws attention as the universal 'clock' by the most significant aspect of its metamorphoses, their dual character – the fact that the moon grows to be a full disk, and during the second it gradually declines to complete disappearance. The 'turning point' usually arrives on the fourteenth day from its appearance, and after another 14 days it disappears. Identification of this important number – 14 – in the form of notches, incisions, and so forth may have occurred long before any abstract concept about it emerged.

If Frolov is correct that the observation of the moon was a central concern of Upper Palaeolithic peoples, then their recording of its cycle in the concrete form of a calendar should be present among surviving artefacts from the period. He cites a notched trowel from the site of Avdeevo in Russia as evidence of the conscious recording of the lunar cycle. This artefact has 4 + 8 = 12 notches on its haft, 8 + 7 = 15 on the head, and a further two on the joint linking the haft and head; the total being 29, the number of days in the lunar month. One might object to the number of notches on this artefact being interpreted as a record of the lunar month, as no explanation is given as to why the 4-8-8-7-2 division was chosen by its maker. There is nothing in the lunar cycle to suggest a subdividing of the 29 days in such an arrangement, and the notches on the trowel could simply have fortuitously added up to 29. From this perspective it might be argued that the notches had no significance whatsoever to their maker, or if they did then they may have represented something other than the lunar month.

Yet Frolov is not alone in interpreting kindred artefacts as expressing lunar notations. In 1963 a limestone object was recovered from an Upper Palaeolithic site near Bodrogkeresztur in northern Hungary that may also be evidence for the existence of the lunar calendar in the Old Stone Age. The object is $56 \times 56 \times 17$ millimetres and has been variously described by the archaeologist László Vértes as 'shaped like a halfmoon or horseshoe', and having a 'lunar or solar shape'. The object (shown in *Figure 23*) was originally interpreted by Vértes as a representation of the uterus, but after having become acquainted with the work of the American researcher Alexander Marshack (for which see below, pages 100–106), he thought it could also possibly be a lunar calendar. As can be seen from the illustration of this object, it is artificially notched along its edges. Although it is not possible to say with total certainty which way up the object is supposed to be viewed, it seems most likely that Vértes has placed it in the most plausible way. If so then one may see (as it faces one) that at the top of the object there are two almost vertical carved lines; to the right there are 11 notches, to the left 12 notches and near the bottom is a single horizontal line. In the diagram the interpretation of the object as indicating the lunar cycle is shown.

In this tentative reconstruction the line at the top left is to be seen as a

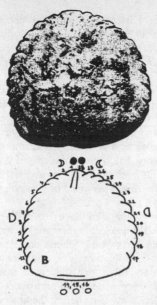

Figure 23

sign for the new moon (hence the depiction of a black disc and the number 1 that have been put in by Vértes to show this); the 12 notches on the left-hand side of the object each represent one of the days of the crescent moon (13 days including the new moon), the bottom, which has no notches, may nevertheless indicate the three days of the full moon (days 14, 15 and 16 in the lunar cycle) by means of the horizontal line; and the 11 notches on the right-hand side the period of the waning moon (days 17 to 27); and finally the right-hand line at the top would be the last day of the cycle (day 28) when the moon disappears once more. Ingenious and interesting as this interpretation is, it is also open to other interpretations, particularly as it lacks separate notched marks for days 14, 15 and 16 which, in the reconstruction, are all symbolised by the single horizontal line. For Vértes the viewing of the object as a uterus and as a lunar calendar were by no means mutually exclusive interpretations – far from it, for he saw that the two could be plausibly linked to the notion of a Great Goddess whose symbolic attributes may include both the moon and the uterus.

Another researcher into the archaic roots of astronomical observation, or palaeoastronomy as he calls it, is Vitaliy Larichev, who, following in the footsteps of Frolov, has interpreted the Upper Palaeolithic inhabitants of Siberia to have attained a not inconsiderable grasp of scientific principles. He has also characterised his own research into this arcane subject in an article he wrote with two colleagues a few years ago, in which he announces in no uncertain terms: 'Larichev's work has penetrated the innermost meaning of ancient art objects and revealed them as carriers of information.' Among such information he notes the symbolic artefacts from the central Siberian site of Mal'ta as recording – amongst other things – both the solar year and the phases of the moon, and believes that the calendars of the Upper Palaeolithic inhabitants of Mal'ta, some 24,000 years ago, were particularly made as a means to predict solar and lunar eclipses. He also believes that an object found at the even earlier site of Malaya Siya in western Siberia (about 34,000 BP) may be an extremely archaic but nevertheless rather complex astronomical instrument of the calendrical type, and that a mammoth-ivory baton from the site of Achinsk, another early Upper Palaeolithic site, is engraved not for decorative purposes but as 'a numerical pictogram reflecting three years of a complex lunar-solar calendar system'.

Frolov has noted a curious parallel that exists in diverse parts of the world concerning the Pleiades. He notes that they are known as the Seven Sisters by native peoples in North America, Siberia and Australia, a fact that he believes to be impossible to be merely a coincidence. Yet if it is not a coincidence, then the only other explanation is that it must point to a common heritage for this very widespread occurrence. If this were the case then it must have its origin some time before the first occupation of the New World (which most archaeologists would place at around 12,000 BP) and even before the first peopling of the Australian continent (about 40,000 BP). Therefore, not only the actual observation of the Pleiades but also the descriptive name of these stars as the Seven Sisters must date back more than 40,000 years. Most researchers tend to ignore anomalous findings such as these, as they simply do not fit well with generally accepted views. If the occurrence of the image of the Seven Sisters in three disparate parts of the world is not a coincidence, then one is left with the alternative described by Frolov – namely a tradition of communicable knowledge of the heavens that has existed for over 40,000 years, since a time roughly coinciding with the beginning of the Upper Palaeolithic. This is something which is extremely awkward for most widely accepted views of the history of knowledge and science – in short, it is far, far too early for most people to accept. Nevertheless, it would certainly provide a period of formation out of which Larichev's reading of certain objects of early Upper Palaeolithic age as comparatively advanced examples of calendars based on both solar and lunar observations could have developed.

Of all researchers into the roots of astronomical observation and the manifestations of such knowledge in notational systems in the Upper Palaeolithic, it is the figure of Alexander Marshack that stands out, both for his continual occupation with these matters for some 35 years and for his numerous insights into the subject. In fact many of the musings of Frolov, Larichev, Vértes and other scholars who have sought the origins of astronomical observations in the Palaeolithic age have relied heavily on

Marshack's discoveries and analyses of numerous artefacts from this remote time. The story of Marshack's search for what he called the roots of civilisation began in 1962 when, as a science writer, he was commissioned to write a book with the NASA scientist Robert Jastrow to explain the background to the projected moon-landing that was on the cards at this time. Part of the book was to provide an account of the historical background that had made this possible. Whilst researching the prehistoric roots of science and civilisation, Marshack began to feel a deep sense of exasperation and found existing explanations on the subject inadequate. He felt there was some kind of missing link: not in the biological sense but rather in a cultural sense. As he puts it:

> Searching through the historical record for the origins of the evolved civilisations, I was disturbed by the series of 'suddenlies'. Science, that is, formal science, had begun 'suddenly' with the Greeks; in a less philosophically coherent way, bits of near-science, mathematics and astronomy, had appeared 'suddenly' among the Mesopotamians, the Egyptians, the early Chinese, and much later, in the Americas; civilisation itself had apparently begun 'suddenly' in the great arc of the Fertile Crescent in the Middle East; writing (history begins with writing) had appeared 'suddenly' with the cuneiform of Mesopotamia and the hieroglyphs of Egypt; agriculture, the economic base of all the evolved civilisations, had apparently begun 'suddenly' some ten thousand years ago with a relatively short period of incipience, or near-agriculture, leading into it; the calendar had begun 'suddenly' with agriculture; art and decoration had begun 'suddenly' some thirty or forty thousand years ago during the Ice Age, apparently at the point during which modern *Homo sapiens* man – as one theory has it – walked into Europe to displace Neanderthal man.

Marshack's growing dissatisfaction with the standard accounts of ancient history and prehistory led him to seek out a more plausible account of events. The turning point came when he read a short article that described an engraved bone found at the Mesolithic site of Ishango in the Congo, dated to about 8500 BP. He believed that the engraved marks on the bone were not only intentional but had a very specific meaning and purpose. Years before he had any interest in prehistory, he had been struck by the fact that the various different kinds of research undertaken by modern scientists – whether oceanography, meteorology, astronomy or even the space programmes themselves – were 'time-factored' and required the recording of measurements and other data over time. On a hunch he felt that this modest artefact from the Stone Age might also be 'time-factored'; he compared the grouping of marks on the bone with the lunar phases and found them to be a pretty good match. He soon realised that this bone artefact could be evidence for a lunar notation that was non-arithmetical, in that the days were not counted and calculated arithmetically but were recognised by their positioning into sets with breaks or special markings which indicated phases of the lunar cycle.

Marshack was hooked. In fact, he was hooked for life. He gave up the pioneers of the Space Age, who were only a few years from landing on the moon, for the pioneers of the Stone Age, who thousands upon thousands of years ago were keeping permanent records of their own lunar observations by accumulating marks or notches over a period of time. By doing so he became a pioneer in the study of the cognitive aspects of prehistory and has undoubtedly contributed more to our understanding of the marked bones and other related artefacts from the Upper Palaeolithic and Mesolithic periods then any other individual to date. He has effectively, to the satisfaction of many prehistorians – though by no means all – established the existence of lunar notations throughout the Upper Palaeolithic from the earliest period, the Aurignacian, onward during the course of studying literally thousands of artefacts. Contrary to what some of his critics have said, he has not seen lunar calendars in all or even most of the marked artefacts he has studied, and he has rejected many possible examples of such 'time-factored' markings as the result of close examination of the artefacts in question.

We have seen earlier in this chapter that people do not develop numerical systems of any complexity without a need to do so, and this also applies to lunar notation. Marshack does not see this kind of activity in the Stone Age as resulting from an extremely precocious and dispassionate intellectual interest in the workings of nature. If it is not a basic prototype for modern scientific practice, then what is it? For Marshack the need, whilst not purely driven by economic and practical concerns, was nevertheless intimately connected with the observation of the seasonal changes in climate, vegetation, the seasonal migration of herds and the biological cycles – particularly those relating to menstruation and pregnancy. But the 'time-factored' interests of prehistoric people manifested in these systems of notation were also as much about the mythological and ritual aspects of culture. Anthropologists have made it clear that the various spheres of life which are for us today so clearly separated (such as the economic, the religious and the scientific) are not divided in anything like the same way among many tribal societies. The same was undoubtedly true of the societies of the Upper Palaeolithic. The cultural body of knowledge in such small-scale societies has not been dismembered into the art world, the business world, the academic world and so on. Their body of knowledge was still intact and holistic; art, science, religion and economic concerns were intimately and inexorably linked together. Marshack has proposed that the observation of the moon was the basis of the 'time-factored' concerns of the peoples of the Upper Palaeolithic and that it represented the fundamental framework upon which their various practical and ideological concerns were constructed.

The most meticulous examination of such objects undertaken by Marshack has been his decipherment of the Taï plaque, which he has described as 'in many ways the most complex single artifact to come from the European Upper Palaeolithic'. Although it is less than nine centimetres long, this engraved bone plaque is clearly of a sophisticated nature, as is attested by both the complexity of the arrangement of the markings on it and their sheer number. This artefact was discovered in 1969 at the Grotte du Taï in western France and dates to the end of the Upper Palaeolithic some 10,000 or 11,000 years ago. Marshack spent about twenty years on and off trying to break its code, using both a microscope and the naked eye. He also wrote that his final success could not have been attained without a study 'of all the Upper Palaeolithic symboling traditions, but a study also of the known traditions of notation found among the world's preliterate cultures'. To give a detailed blow-by-blow account of the decipherment of the markings on this object is simply not possible here. The reader who wishes to have such an account is referred to Marshack's extended exposition of it over the course of 35 pages in an article published in 1991 (see Bibliography). For our present purposes it is sufficient simply to describe the most essential features of the object and Marshack's conclusions concerning its fundamental significance.

The plaque is part of a rib bone that appears to have been salvaged and cut into a suitable size and shape for the purposes of its Ice Age engraver. Its portability would have been a key factor in its selection, for it is probable that its owner belonged to a mobile population. Marshack's previous studies of other comparable, if simpler, artefacts from the Upper Palaeolithic assisted him in discovering how the sequence of markings should be 'read' in this particular instance. He saw it as an instance of a mode of marking known as the serpentine or 'boustrophedon' type, which had been one of the methods used since Aurignacian times; that is, from the onset of the Upper Palaeolithic. The term boustrophedon is usually used with reference to writing, and describes a system which involves writing one line with the script to be read from one direction (left to right or right to left) and then writing the next in the opposite direction and continuing this alternation throughout the text. Applying this to the Taï plaque and other systems of notation that are not writing, it means that the sequence of lines was made starting from either right to left or left to right and then alternated, as in the case of writing. In some instances the lines would not be separated from one another but would 'turn the corner', thus making a serpentine pattern. He also notes the possibility that the serpentine arrangement of such systems of notation may have been, 'unwittingly, an abstracted image of periodicity'. He found that the sequence of sets of marks on the plaque was clearly the result of lunar observations.

He had previously obtained lunar notations from a number of other artefacts, so this feature of the Taï plaque was not in itself remarkable. What was striking was that during the course of his analysis of this lunar notation he noticed a certain break in the sequence of notation that occurred after the recording of six lunar months. This break was clearly due to the observation of some time factor external to the lunar recording itself. Although it is possible that some stellar observation may have been the cause of this break, the most obvious answer was that it marked the occurrence of the solstices or the equinoxes. Marshack concluded that barring the unlikely alternative of stellar observation, this kind of break in the sequence pointed clearly to the observation of the sun. Thus the especial significance of the plaque was that its notation combined observational data from both the sun and the moon and thus was a solar/lunar calendar. The complex nature of the Taï bone plaque is for Marshack the outcome of a body of knowledge that can be traced back to the Aurignacian period and the comparatively simple lunar notations that he had identified on artefacts from this period. In this developmental view of Upper Palaeolithic astronomical notations, he parts company with Larichev, who, as I mentioned above, has claimed that the lunar/solar calendar is to be found on Siberian artefacts dating from 22,000 or even 34,000 years ago to the early Upper Palaeolithic period. Whether Larichev's claims gain wider acceptance remains to be seen but for the moment Marshack's propositions are more generally persuasive.

The significance of the Taï notation lies not only in the fact that it demonstrates the developed nature of the human observation of the sun and the moon in the Upper Palaeolithic era. It also fills the vacuum that previously seemed to have existed before the apparently sudden development of astronomical observations in the Neolithic period in north-western Europe, epitomised by the alignment of megalithic monuments such as Stonehenge. With the Taï plaque there is no longer a need to seek the roots of Neolithic astronomical knowledge in Europe from the more 'advanced' centres of culture in the Near East. This knowledge could easily have grown organically out of native tradition, as Marshack believes:

That a non-arithmetical observational astronomical skill and lore existed in Europe before the Neolithic suggests that adoption of a farming way of life was prepared for or assisted by the seasonal economic and ritual calendar of the Palaeolithic hunter-gatherers. If so, the astronomical knowledge of the European Neolithic may have been essentially indigenous and need not have entered that continent with the farming technologies and seeds of the Near East. The farming technologies may have been assimilated into an already extant European lore of the seasons and the year. There is a wealth of European folklore suggesting retention of an indigenous pre-Neolithic mythology. The alignment astronomies of the West European Neolithic suggest that an observational lunar/solar calendar such as that represented in the Taï notation may have provided the base for the observational lunar/solar 'calendars' of the megalithic cultures. This European observational astronomy seems not to have been derived from the Near East or Egypt, but from the Palaeolithic tradition documented by the Taï notation.

The idea that the Neolithic and early Bronze Age megalithic monuments of north-western Europe had drawn direct or indirect inspiration from more 'evolved' civilisations received a body blow when the development of radiocarbon dating showed that many of the megaliths were in fact much older than they had previously been thought to be. The most famous of the astronomically aligned sites in the British Isles, such as Stonehenge and Newgrange in Ireland, were once thought to have been due to influences from more 'civilised' cultures from the East, and the idea that the barbarians of northern Europe could have created such monuments on their own initiative was, to some, unthinkable. Since these two sites are known to be about 5,000 years old, it is now clear that the megalithic cultures of north-western Europe relied more on their own indigenous knowledge than on imported 'civilised' notions. What Marshack has added to this picture of the independent development of such architectural traditions is to show that the astronomical knowledge which the megaliths embody is also an outgrowth of native culture. He has thus traced the route of this scientific knowledge not from East to West as earlier thinkers had suggested, but from its roots in the Old Stone Age traditions of the West. In the next chapter, the clues that lead back to another ancient path of scientific thinking reveal that the early hunters relied on more than just a sharp spear and a strong arm to pursue their quarry.

Chapter 7

From Footprints to Fingerprints

In an illuminating essay entitled 'Clues: Roots of an Evidential Paradigm', the renowned Italian historian Carlo Ginzburg describes a very important but little-understood set of methods used by investigators in a number of fields of enquiry. This analytical approach is undertaken by the close and careful study of seemingly trivial or unimportant details which actually turn out to be of great importance. This is best illustrated by the comparatively recent examples that first set Ginzburg on the trail of this kind of thinking, which can be traced back to remote times. He begins with the method of identifying paintings developed by the Italian Giovanni Morelli in the 1870s. Morelli realised that many paintings in art museums were unsigned and difficult to attribute to a particular artist with a sufficient degree of certainty by using conventional methods. In an artistic world bedevilled by forgeries, copies and misidentifications, Morelli believed that concentrating on the most obvious and illustrious aspects of any given painting was not the way to exorcise the spirit of confusion. He rather proposed that the apparently trivial details of paintings, such as the way in which earlobes, fingernails, aureoles and so on were executed, was a means of bypassing the more visually striking properties of a painting which a copyist could fairly easily imitate. By concentrating on these minor aspects of paintings, largely ignored by the critics as well as the copyists, Morelli believed the identity of the painter could be revealed.

Ginzburg notes a similar approach in the sphere of crime detection, epitomised in the methods employed by Sherlock Holmes in the fictional writings of Sir Arthur Conan Doyle. Holmes, unlike the less gifted and more pedestrian thinker Dr Watson, pursues unusual analyses of seemingly irrelevant details such as cigarette ash in tracking down his quarry. As Ginzburg points out, the similarity between the detective's approach and that of Morelli is transparent in one particular story, called *The Cardboard Box*. In this case Holmes is able to deduce from two severed ears sent to an unfortunate woman that their shape indicates not only the sex of the victim but also her close blood relationship to the woman who received them. The third example of the use of such techniques is drawn from the work of someone even more famous than Sherlock Holmes, namely Sigmund Freud. Ginzburg points out that the parallels between Morelli and Freud were made by the great psychologist himself, who wrote: 'it seems to me that his

[Morelli's] method of inquiry is closely related to the technique of psycho-analysis. It, too, is accustomed to divine secret and concealed things from unconsidered or unnoticed details, from the rubbish heap, as it were, of our observations.'

By his own tracking down of these parallels between the investigation of pictures by Morelli, clues by Holmes and symptoms by Freud, Ginzburg proves himself to be a hunter of some prowess in the dense undergrowth of the historical archives. He notes the necessity of such trains of thought in the study of both medicine and history (we may add prehistory) as well as other disciplines that inevitably have to deal with individual cases, events and documents. He also sees a particularly striking example of the practical application of such methods in the development of the fingerprinting technique, in which an apparently trivial aspect of a person's physiognomy turns out to be the best means of identifying its owner. In 1823 J.E. Purkyné became the first man to observe that the fingerprints of each individual were unique to that person, yet it was not until towards the end of the century that fingerprints began to be used as a means of identification. Ginzburg sees these methods of analysis as having a remote origin in prehistory. As he puts it:

Man has been a hunter for thousands of years. In the course of countless chases he learned to reconstruct the shapes and movements of his invisible prey from tracks on the ground, broken branches, excrement, tufts of hair, entangled feathers, stagnating odors. He learned to sniff out, record, interpret, and classify such infinitesimal traces as trails of spittle. He learned how to execute complex mental operations with lightning speed, in the depth of a forest or in a prairie with its hidden dangers.

And again:

Behind this presumptive or divinatory paradigm we perceive what may be the oldest act in the intellectual history of the human race: the hunter squatting on the ground, studying the tracks of his quarry.

Ginzburg does not pursue the prehistoric trail himself, but his discovery of its traces in important aspects of modern thought is highly illuminating. The use of tracking techniques is essential for successful hunting strategies. Whilst there are some prehistorians who would still argue that hunting in any systematic sense only began in the time of behaviourally modern humans, about 40,000 years ago, and that earlier man was merely a lowly scavenger, it now seems clear that there is strong evidence for hunting with throwing spears hundreds of thousands of years ago, long before even the Neanderthals (for details of this, see Chapter 11). We may say that the intellectual activities involved in tracking must indeed be among the first use of such faculties, and may be, as has been suggested by the anthropologist Louis Liebenberg, the ultimate origin of scientific reasoning. If we look at the tracking techniques of the San, the modern hunter-gatherer peoples of the Kalahari desert in southern Africa (popularly known as the Bushmen), it becomes clear how striking are the similarities with the conjectural paradigm explored by Ginzburg in its modern Western manifestations. Liebenberg, who makes his observations without knowledge of Ginzburg's research, makes the same basic points about the thought processes of the Kalahari hunters as Ginzburg does about Morelli, Sherlock Holmes, Freud and the invention of fingerprinting:

> The modern tracker [of the Kalahari region] creates imaginative reconstructions to explain what the animals were doing, and on this basis makes novel predictions in unique circumstances. Speculative tracking involves a continuous process of conjecture and refutation to deal with complex, dynamic, ever-changing variables.

Thus tracking, like psychology and historical research, is concerned with individual (unique) circumstances which require conjectures to be success-ful in the pursuit of their respective quarries. Liebenberg also believes that it is in the art of tracking that we may find the wellsprings of the scientific quest. He sees the evolution of the art of tracking as having three stages. The first, which he calls simple tracking, is the following of animal footprints in ideal conditions in which the trail is easy to discern. The second, systematic tracking, involves essentially the same kind of thought processes but requires the collecting of data in more difficult conditions. It is thus a more sophisticated version of simple tracking. The third, speculative tracking, requires the hunter to develop a working hypothesis using various kinds of data – knowledge of the animal's behaviour patterns, the terrain and so on. From this hypothetical reconstruction of the activities of the animal, the hunter then, with this in mind, looks for the most likely place to find tracks and other signs. At first consideration it may seem rather unlikely that traditional hunter-gatherers really use scientific or even quasi-scientific methods in their pursuit of game. We are accustomed to thinking of scientific methods and intellectual analysis as belonging to the laboratory, the classroom and the library rather than the activities of hunters. Hunting is usually considered primarily as a practical activity in which intellectual speculations would be not only out of place but perhaps even an actual liability. Yet anthropologists who have worked closely among the hunters of the Kalahari and elsewhere have found that hunting is not simply an instinctual practice but also involves considerable and occasionally great learning and intellectual insight; the so-called modern 'primitive' – and, by extension, prehistoric 'primitive' – cannot be seen as an unthinking 'savage'. The subject that the hunter studies is not atomic particles, historical documents, crimes or paintings, but spoor and other signs of animal activities. An accurate and detailed knowledge of spoor is essential for hunting peoples. Although the word is often used to refer simply to footprints, it also has a wider meaning, which refers to signs left by animals on the ground and in vegetation. As Liebenberg explains, there are numerous sources of data used by hunters, such as:

> Scent, urine and faeces, saliva, pellets, feeding signs, vocal and other auditory signs, visual signs, incidental signs, paths, homes and shelters. Spoor are not confined to living creatures. Leaves and twigs rolling in the wind, long grass sweeping the ground or dislodged stones rolling down a steep slope leave their distinctive spoor. Markings left by implements, weapons or objects may indicate the activities of the persons who used them, and vehicles also leave tracks.

Thus we can see that the hunter has a number of kinds of data that must be assessed and considered in conjunction. Just as the detective pursues a criminal by assessing the various clues and evidence available to him or her, so too does the hunter in pursuit of game. This parallel between hunting and criminal detection becomes even more apparent when we consider the art of tracking in more detail. The fluent reader of the signs of spoor rarely has a simple set of complete footprints to follow and has to be able to identify the species on the basis of partial prints; grass bent by the passing animal may also provide vital information on the direction of its movement. As I have shown a number of times in this book, archaeologists often indulge in research of an experimental nature. Experimental archaeology typically involves experiments in the reproduction of prehistoric or ancient artefacts or technological processes in order to better understand the objects that they have unearthed. Hunters may also conduct their own simple 'on the hoof' experiments. Broken twigs and branches found on the ground during tracking may or may not be recent. To clarify to himself whether the break is recent or not, a hunter will sometimes pick up similar twigs and break them himself in order to make a comparison. An old break will appear weathered and dull in contrast to the freshly made break. If the break in the twig or branch is fresh, it can then be considered as a relevant piece of data in the hunt. Various additional factors, such as the season, the weather, the terrain, soil type and conditions and so on, all have to be considered in the hunter's assessment of the spoor trail. Hunters of the Kalahari desert are able to discern, even in loose sand, the spoor of numerous creatures, ranging from beetles and millipedes (see *Figure 24*) to the snake and the mongoose. They are even able to distinguish different species of mongoose by the spoor alone.

The study of animal faeces can also provide important information to the hunter, indicating not only the species that deposited it but also details of diet, how long ago the animal passed by and, in some cases, even whether the animal itself was young or old, something which can be established by the size of the faeces. Urine traces also provide similar clues. For example,

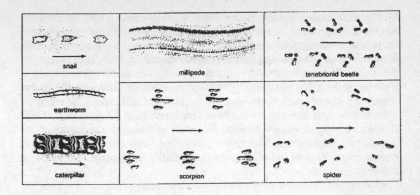

Figure 24

an antelope urinates with its back legs straddled, and knowledge of this fact allows a hunter to deduce exactly where the animal was standing at the time. The voiding of urine by the male and the female antelope produces a urine patch in a different spatial relationship to the footprints of each. In the case of the male, the urine will be between the fore feet and hind feet tracks, that of the female between or behind the hind feet. Although the human sense of smell is vastly inferior to that of other predatory beasts, an experienced hunter among the San is nevertheless able to distinguish the different odours of a variety of animals, including elephant, zebra, wildebeest and buffalo, if sufficiently close. Blood traces left by an animal wounded by either a hunter or a rival carnivore are also a source of useful information. If the blood is clear and contains bubbles of air, it indicates a wound to the lungs; if it has come from a stomach wound the blood will contain stomach contents. The occurrence of blood stains on vegetation can help to establish the height of the wound on the animal's body. A trail which consists of small spots with relatively large gaps in between the blood spilt on the ground indicates that the wound well may be superficial and that it has not impaired the ability of the animal to run. In contrast, heavy loss of blood and the weakening of the quarry is indicated by larger spots that are also closer together. Saliva and water splash marks are indications of a very recent presence of an animal, as both are soon evaporated.

The hunter's art of tracking is by no means limited to putting together a mental model of the quarry's identity and position purely on the basis of visual signs. Various sounds can also lead the hunter to his prey. The presence of a rival from the animal kingdom can be announced to the hunter by a bird that may cry out a warning, but this can also alert the hunter's own quarry to his presence. The speed and size of an animal can sometimes be indicated by various noises, such as the disturbance of vegetation, the breaking of branches or twigs and the splashing of water. Even the absence

111

of expected sounds – such as the chirping of crickets – can be a useful source of information. Many animals have their own network of paths and trails which often branch out in numerous directions from water holes and favoured feeding zones, and naturally a knowledge of these is important to the hunter. The various abodes of animals, such as nests, burrows and drays, likewise receive considerable attention from trackers. The identity of the owner of a burrow may be deduced by the position and size of the entrance hole, and the way in which the earth was removed from it during its making. If it is still being used, this is often indicated by tracks or droppings in the entrance or in the immediate vicinity. If it has been abandoned, then the entrance will often be full of dead leaves, partially overgrown or covered in cobwebs.

Many more examples of the data sources of hunters could be listed, but those given above show just how many factors may be combined in the developing of a hypothesis concerning the quarry. It is easy to see how hunting techniques are comparable to the methods described in both fictional and factual accounts of detective work, for it is the same kind of mental processes that are being used in both hunting and crime-solving. Nicolas Blurton Jones and Melvin J. Konner, who undertook fieldwork together among the !Kung hunters in 1970 as members of the Harvard Kalahari Research Group, also see parallels between Western intellectual pursuits and !Kung hunting:

> Such an intellective process is familiar to us from detective stories and indeed also from science itself. Evidently it is a basic feature of human mental life. It would be surprising indeed if repeated activation of hypotheses, trying them out against new data, integrating them with previously known facts, and rejecting ones which do not stand up, were habits of mind peculiar to western scientists and detectives. !Kung behavior indicates that, on the contrary, the very way of life for which the human brain evolved required them. That they are brought to impressive fruition by the technology of scientists and the leisure of novelists should not be allowed to persuade us that we invented them. Man is the only hunting mammal with so rudimentary a sense of smell, that he could only have come to successful hunting through intellectual evolution.

The vast majority of the modern human population have long since abandoned the hunter-gatherer lifestyle for other economic systems based upon agriculture and industry. Hunting no longer holds pride of place as the means of obtaining food and the other necessities of life. The practice of hunting nevertheless continues as a leisure pursuit and as a means of controlling vermin. Many sporting hunters justify their pastime as an antidote to the restrictions and repression of civilisation, for it provides a way of getting back to nature, to the primordial human state. Many people find the whole notion of hunting for pleasure barbaric and mindless, an attitude most markedly found among urban populations. It must be said that, at least in many cases, they are correct. Big businessmen who go to places like Alaska in order to 'waste' game animals from the comfort of helicopters and using high-powered rifles are hardly 'going back to nature' or facing a real test of their hunting skills. Their motives are clearly of a more visceral nature, the killing of the animal being simply a macho device for boardroom bragging. Such businessmen are the ignoble heirs of the colonial hunters who, using the hunting skills of their ethnic subjects, would shoot tigers, lions and other big game simply to collect trophies. The rather vacuous rituals of fox-hunting among the British upper classes show that the age-old link between royalty and hunting – which can be traced back to the ancient societies of Egypt, Mesopotamia and Iran – is still maintained even today. Clearly the distinction between most modern hunting and that of traditional hunter-gatherers is based on the different underlying motives – hunting for food and hunting for pleasure.

Another distinction that is often made between the two kinds of hunting is that traditional hunters are often seen as having a greater degree of respect for their quarry and even a pang of regret at having to kill animals at all. Liebenberg asked one of the San hunters from whom he was learning traditional tracking techniques how he felt about the animals he killed. The hunter replied that he did have a certain sympathy for them but he nevertheless had to kill them in order to eat. The mature antelope inspired less sympathy than a younger one, since the former 'knew the game' and was aware of the dangers that a hunter posed. The juvenile antelope, on the other hand, was callow and naïve and if one was caught in the hunter's snare it made him feel sad. Even a beetle with a broken leg could inspire pity, but he did not feel sorrow when killing a scorpion, which would happily sting him if it were given the opportunity. Liebenberg also quotes another case of sympathy for the quarry, this time from *The Forest People*, Colin Turnbull's account of the Pygmies who live in Central Africa; one can add numerous cases of empathy and even reverence for animals among the hunting peoples of northern Canada and elsewhere. Based on such attitudes among traditional hunting societies, it has been proposed that wanton cruelty to animals belongs to a later phase of human development and that, despite the necessity of killing for food, hunter-gatherers lived in mutual respect and harmony with their animal neighbours. Attractive as this portrayal of a hunter-gatherer Eden is, there are also accounts of cruelty in such communities. Even in Turnbull's book, which Liebenberg cites as supporting evidence that hunters such as the Pygmies were gentle and compassionate towards animals, there are accounts of them laughing and kicking a wounded animal, and also singeing the feathers off live birds.

Whilst those who actually practise hunting today represent a very small minority, the use of hunting metaphors and techniques continues unabated, even in the highly unnatural environment of the modern sprawling metropolis. Humans have spent over 90 per cent of their existence as hunters, so it is hardly surprising that such metaphors imbue our collective psyche – the Neolithic onwards represents merely the tip of the iceberg. For example, we have seen how the detective's *modus operandi* draws on a set of procedures and insights that ultimately derive from hunting, and this is also reflected in phrases such as 'tracking down a killer', 'setting a trap', 'using a decoy', 'covering his tracks' and so forth. Yet these are not simply metaphors but concrete procedures by which the detective can actually track down his or her human quarry. Whilst the hunter stalks his animal prey at the water hole, detectives will seek out criminals at their 'watering holes' (bars); as the hunter may gain vital information from a bird's call that may indicate the presence of the quarry, so an informant may likewise provide the detective with such leads.

The historian's problem is often to sift through a vast amount of textual information and cut his or her way through the dense undergrowth of archival information in order to hunt down and gather up the required information. By contrast the prehistoric archaeologist has already had his or her database scythed away by the ravages of time and somehow has to reconstruct the events of the remote past using the often infinitesimal remnants of archaic cultures. Whilst historians may sometimes make significant breakthroughs in their work by unearthing apparently insignificant and minor facts which may lead them to important discoveries, the prehistoric archaeologist is often *obliged* to pursue such a methodology simply because the traces of human activity left after thousands of years are usually of a minor and incomplete nature. Thus for the historian the use of such methods is often down to choice, whilst for the prehistorian it is usually a matter of necessity. Freud, as quoted above, used the metaphor of 'the rubbish heap of our observations' to describe the source of vital clues in psychoanalysis. The archaeologist pursues a similar task but on a literal level. Ancient and prehistoric rubbish dumps may have been merely places for earlier peoples to throw away what they considered was waste, but for the archaeologist, rubbish may be as important as it is to the detective or the intelligence officer. This cultural dross provides clues into all kinds of aspects of cultural life. As the hunter may pause to inspect animal faeces on the trail for clues, so likewise the archaeologist may analyse coprolites (fossilised faeces) for dietary data.

Yet it is not only the methods used by the prehistorian that mirror some of the intellectual and practical procedures of archaic hunting techniques. The prehistory era itself is still generally considered as of marginal importance in comparison with the historic era. Most accounts of the human story pay little attention to the prehistoric phase, it being seen as little more than a prelude to the real action and excitement represented by history and the large-scale significant events of the great civilisations. To my mind such accounts can be nothing but gross distortions. If we instead make use of the kinds of methods found in the works of Morelli, Conan Doyle and Freud and look at the human story in this light, then a different view emerges. We find the apparently insignificant parts of the human story – namely the prehistoric era – to be the key to the underlying reality. In short, we find that historical civilisations do not represent a quantum leap away from the savage, muddled and ignorant societies of the Stone Age, but rather an elaboration, an accumulation and magnification of the insights and creations of their prehistoric ancestors.

Anthropologists have been keen to point out that there is no such thing as a natural society, since even the smallest of hunting bands have their own cultural lives. Nevertheless, members of such societies have a more direct daily contact with the natural world around them than their counterparts in the urban world. Thus their use of the intellectual processes discussed in this chapter is orientated towards hunting procedures such as tracking rather than activities such as historical and archival work, building psychological profiles of serial killers or tracking by radar. Yet the same fundamental thought processes are at work in both the Kalahari and at Harvard, as Blurton Jones and Konner make clear when assessing the knowledge of animal behaviour (ethology) and tracking among the !Kung:

> The accuracy of observation, the patience, and the experiences of wildlife they have had and appreciate are enviable. The sheer, elegant logic of deductions from tracks would satiate the most avid crossword fan or reader of detective stories. The objectivity is also enviable to scientists who believe that they can identify it and that the progress of science is totally dependent upon it. Even the poor theorisation of our !Kung left one uneasy; their 'errors,' the errors of 'Stone Age savages,' are exactly those still made today by many highly educated western scientists . . . We have gained little or nothing in ability or intellectual brilliance since the Stone Age; our gains have all been in the accumulation of records of our intellectual achievements. We climb on each other's backs; we know more and understand more, but our intellects are no better. It is an error to equate the documented history of intellectual achievement with a history of intellect. It is an error to assume that changes in about 7,000 years of urban civilisation represent a final stage in a progress which can be extrapolated downwards into our preurban past. Just as primitive life no longer can be characterised as nasty, brutish, and short, no longer can it be characterised as stupid, ignorant, or superstition-dominated.

Chapter 8

Under the Knife

We have already seen how hunting peoples have an acute understanding and a detailed knowledge of the world around them, and, as we know from recent tribal cultures, the daily experience of butchery and the gathering of plants combined a basic knowledge of anatomy with the use of medicinal plants. Nevertheless, it seems that whilst the treatment of wounds, blood-letting and bone-setting are common in tribal cultures, other surgical operations for medical purposes are apparently rather rare. In many cases this does not appear to be due to any lack of ability to wield the knife in a competent way. This is made clear by the various ritual mutilations or alterations of the human body performed almost as a matter of course in numerous small-scale societies throughout the world. The amputation of one or more fingers in initiation ceremonies and other rituals is widely reported from Africa, Oceania and North and South America, and the discovery of the outlines of hands with parts of fingers missing in some of the decorated caves of France and Spain has been interpreted as evidence of this practice in Upper Palaeolithic times, although the more prosaic suggestion that the loss of fingers may be due to frostbite or gangrene has also been put forward as a plausible explanation.

The performing of circumcision, especially among the Jews and the Muslims, is of course a well-known and ancient practice, but other genital operations of a ritual rather than strictly medical kind are performed in various parts of the world. Another practice is subincision, which involves the opening of the male urethra by a cut from an inch long to, in some cases, an incision extending all the way to the scrotum. Although known elsewhere, it is largely the speciality of the Aboriginal inhabitants of Australia, having traditionally been performed in the context of initiation rites in some three-quarters of the continent. Female initiations in Australia also often involved some equivalent cutting of the genital region, from simply lacerating the hymen to 'female circumcision' or clitoridectomy, the removal of the clitoris. Various theories as to why subincision was practised have been put forward, and the totally unfounded and bizarre notion that it was a contraceptive measure has long been rejected. It seems that in Australia the bleeding of the penis that occurs with subincision was a ritualistic act designed to resemble the natural process of female menstruation, both symbolically to obtain female powers and to be a means of expelling 'bad humours' from the body in a way that menstruation is supposed to do.

Similar symbolic reasons for blood-letting from the penis are also known in parts of New Guinea. Among the few other peoples reported to practise subincision, the Fijians and Tongans both believe it to be a therapeutic measure; whilst its purpose among certain Amazonian peoples of Brazil, at least according to nineteenth-century travellers' accounts, was to provide a means to remove tiny fish that occasionally swam up the urethra of the unsuspecting bather. Such fish may be considered almost as dangerous as the legendary piranha! Monorchy, the removal of a testicle, is a surgical operation practised in recent times both in the South Seas and in various parts of Africa. The now notorious process of female infibulation (the sealing of the female genitalia by stitches or other means of fastening; male infibulation was practised by the Romans to restrain sexual activity) has also been practised, most conspicuously in East Africa.

The practice of scalping, whilst usually associated with the native inhabitants of North America, is also known in other times and places and is mentioned by the first Greek historian Herodotus in the fifth century BC as a practice of the Scythian peoples. It is also not widely known that scalping was performed by white Americans on Indian peoples, and in West Africa. Scalping involves the circular cutting and subsequent removal of the skin over the skull either above or below the ears. It was undertaken on both the living and the dead and there are a considerable number of reports of individuals surviving such an ordeal. Sir Richard Burton, the great linguist and intrepid explorer, who had what can be described as a prurient interest in all forms of mutilation – or a supreme level of scientific detachment – avidly collected reports of the various methods of scalping (often known euphemistically as 'raising the hair') practised by the native North Americans, during his 1860 journey to visit the then-novel communities of Mormons in Salt Lake City. Without providing a similar grisly catalogue it must nevertheless be noted that there is evidence for the prehistoric occurrence of scalping that consists of distinctive circular incision marks on the skull bone. Although there can be a variety of explanations for incisions on the skull, in general those that are round in form specifically indicate that the removal of a circular piece of the scalp has been performed, the incisions representing the penetration of the cutting instrument down on to the surface of the skull bone. Deliberate deformations of the head were practised in numerous parts of the world and, at least in pre-Columbian Chile, were even supplemented by devices for inducing facial alterations.

Clearly mankind has never been ignorant of the diverse ways in which the human body can be changed by the removal or alteration of its parts. Amputations are known to have been performed even in hunter-gatherer societies. Stephen Webb's recently published survey of pathological conditions among prehistoric Aboriginal Australians contains details of skeletons that show clear evidence of amputation having been performed.

He believes the practice on that continent to be a last resort and as such unlikely to have been a regular occurrence. He also quotes a quite remarkable late-nineteenth-century case of an Aboriginal amputation originally made known to Revd H. Wollaston, a colonial surgeon:

> At King George's Sound [Western Australia] Mr Wollaston had a native visitor with only one leg; he had travelled ninety-six miles in that maimed state. On examination, the limb had been severed just below the knee, and charred by fire, while about [5 cm] of calcined bone protruded through the flesh. This bone was removed at once by saw, and a presentable stump was made ... On enquiry the native told him that in a tribal fight a spear had struck his leg and penetrated the bone below the knee ... He and his companions made a fire and dug a hole in the earth sufficiently large to admit his leg, and deep enough to allow the wounded part to be on a level with the surface of the ground. The limb was then surrounded with the live coals or charcoal, and kept replenished until the leg was literally burnt off.

Whether or not all the facts of the case are correct, what is clear is that the amputation certainly took place in the bush and involved cauterisation, that is to say burning or searing. There are also numerous cases of amputations and mutilations of various parts of the body as punishments for transgressions and crimes. Even this rapid survey makes it clear that basic surgical procedures existed in many parts of the world, but that such activities seem to have been often undertaken for purposes other than strictly medical ones.

Of all small-scale societies in the world, the native inhabitants of the Aleutian Islands seem to exhibit the greatest knowledge of human and comparative anatomy. William Laughlin, an anthropologist who has devoted most of his professional life to a deep study of Aleut culture, obtained no fewer than 234 indigenous anatomical terms from a single individual named Mike Lokanin, who was only 37 and not even a native doctor! Other informants fleshed out this list for Laughlin, making a total of over 360 separate names for parts of the body, including the tear ducts, various teeth, throat-tendons, the hairs of the armpit, pericadium (the membrane around the heart), spleen, appendix, isthmus (the area between the genital region and the anus), each of the calf muscles, just to name only a random cross-section of this extraordinary inventory of Aleut anatomical knowledge. It should be noted that unlike anatomists in Europe who were able to develop their system by building on the discoveries and terminology of earlier traditions – most conspicuously those of the Greeks and Romans – the Aleuts' anatomical terminology and knowledge were based on their own independent findings. Bearing in mind this fact, and the small scale of their communities when compared with the countries of Europe, the Aleuts' knowledge is all the more impressive.

There are various sources for such knowledge which explain how the

Aleuts attained such a detailed and sophisticated awareness of the structure of the human body. Firstly, their butchering of various animals led them to make use of mammal tissues for making clothing and containers; such tissues went unnoticed in other parts of the world, which indicates the Aleuts' acute knowledge of the potential uses of the creatures that they hunted. Their knowledge did not just accumulate as a by-product of butchering for the purpose of obtaining food and materials for clothing and other practical needs; they also performed dissections on both humans and animals, particularly sea-otters, on account of their anatomical and behavioural similarities to humans. Their traditional medical practitioners would sometimes dissect sea-otters and other animals in order to study the best means to perform operations on human subjects. It was only because the government of the United States banned the killing of sea-otters in 1911 that the practice of comparative dissection declined among the Aleuts. Autopsies were also undertaken by native practitioners in cases where the cause of death was unclear.

Other manipulations of the human body both during and after life were an integral part of Aleut culture. Over a thousand mummies have been found on the Aleutian Islands. On Kagamil Island in the eastern Aleutians, approximately 234 mummies, comprising individuals of both sexes and all age groups, ranging from babies to octogenarians, were deposited in caves, showing that all kinds of individuals – and not just chiefs or other high-ranking persons – in Aleut society were mummified. Their reason for practising mummification was based on the belief that the powers of the deceased individual were beneficial to the living community and that preservation of the physical body kept these powers intact. The corpses of enemies and other outsiders were treated in exactly the opposite way. By the dismembering of such individuals, their baleful powers were believed to be dissipated and thus rendered harmless to the community. The Russians who eventually gained control over the Aleutian Islands recalled with horror the mutilation of their countrymen. Two contrasting incidents reported by Aleut informants illustrate the fact that dismemberment was as fundamental to their social code as was mummification.

Both stories concern encounters with 'outside men', the name given to individuals who chose to live outside of village life and mainstream society and eke out a living by raiding on their own or in small groups. According to William Laughlin, during the nineteenth century a man named Iliodor Sokolnikoff was confronted with an 'outside man' and, in the resulting fray, killed him. He then proceeded to cut off all the man's limbs, systematically at the joints, and then cast them into the sea. By doing so he was free of the negative effects of the dead man's power which would otherwise have stayed intact in the corpse and troubled Sokolnikoff. The other incident concerned a young girl from Umnak Island out alone digging up orchid roots, who was attacked by another 'outside man' who had been watching her, concealed in the long grass. She struck him hard enough on the head with her whalebone root-digger to kill him, but in her shock and fear she ran home to her village without dismembering the body. Because of this she was said to have suffered from painfully inflamed joints when she was older, an ailment that could have been avoided had she followed the traditional practice of dismembering her assailant.

Although certain surgical operations were performed by Aleut doctors, their medical practices were largely based on a system of therapy which is remarkably similar to fundamental aspects of traditional Chinese medicine. Like the Chinese, the Aleuts believe a great range of disorders can be treated by acupuncture. In the Aleut system two basic procedures are followed in acupuncture. In the first method two marks are made on the surface of the skin above the internal organ requiring treatment, and the skin between these marks is pulled up and then pierced by a lancet, traditionally made of flint. According to the Russian Orthodox bishop Ivan Veniaminov, who spent a number of years among the Aleuts in the early nineteenth century, almost all parts of the body were treated in this way. The second method involves pushing the lancet directly into the skin, a procedure requiring greater skill and one which was said to be highly effective. Laughlin believes it is possible that the Aleuts once had a theory of the circulation of blood and the 'airs' of the body akin to that of the Chinese. This tradition may have died out during the contact period with Europeans. Massage was another therapeutic technique that the Aleuts share with traditional Chinese medicine, and involved the manipulation of the internal organs. Massage was, and still is, mainly performed by women, and its most prominent application has been in midwifery. There are other striking parallels with the East Asian cultural sphere, for example in the fighting techniques of the Aleuts that were used both in the sport of wrestling and in actual combat. According to Laughlin:

> They know that pressure or a sharp blow upon the brachio-radialis muscle in the upper lateral forearm will temporarily paralyze the arm. This spot they designate *cam analii* 'the daylight of the hand/forearm', and they have noticed that dogs and sea-lions when fighting also try to bite each other at that spot. Accounts of a strong man killing three adversaries in one action describe him bashing two men's heads together at the temples while crushing a third man's temples between his knees. Aleut wrestling included various types of stationary encounters in which the adversary was thrown or pulled off balance and employing elementary forms of judo which involve, as with the vulnerability of the brachio-radialis and the temples, an acquaintance with the human physique.

That the Aleut techniques of combat were in part derived from their observations of fighting animals has parallels with the exercise regimes of oriental civilisations. Many of the postures and movements of both the

Chinese and Japanese martial arts (such as tai chi ch'uan, shaolin kung fu and ninjitsu) are explicitly modelled on fighting techniques derived from the animal kingdom. In the Indian tradition of yoga, many of the *asanas*, or body postures, are named after animals (e.g. the cock, the lion, etc.). Furthermore, the parallels between the Aleut and the Chinese therapeutic techniques of massage and, more strikingly, acupuncture, in conjunction with the similarities in fighting techniques, clearly show that this is no mere coincidence. Laughlin thinks it is plausible that the Aleuts and the Alaskan Eskimo groups share a tradition with the Chinese that would have its roots in the prehistoric era and to a common Mongoloid ancestral people who inhabited north-eastern Asia. The origin and development of the oriental martial arts is a subject that has received too little attention and is a fascinating line of research that would be most enlightening. Laughlin's suggestion of a prehistoric origin for the acupuncture/fighting technique complex seems to be lent support by the great popularity of wrestling among the Mongolians and the various peoples of north-eastern Siberia, such as the Kamchadal. It is possible that the martial arts that we tend to associate with the Taoist and Buddhist traditions have their prototypes in prehistoric shamanistic practices. We know that shamans have a special place as medical practitioners in tribal societies. They also go into trances in which they imitate both the noises and actions of animals and thus provide a prototype for the later imitation of animal movements by martial artists. The parallels between the Chinese and Aleut systems of combat, medicine and anatomy, if they are indeed drawn from a common cultural source, not only indicate the time depth involved in such traditions in this region of the world, but also provide indirect information on the knowledge of analogous practices in prehistoric times.

Another area of the world which has a remarkable set of medical practices is East Africa, and it is particularly significant for its surgical operations. The native doctors of the region are so adept in the treatment of wounds that they have even been reported to be able successfully to stitch damaged intestines that have been torn open by spears and arrows. They also practise a particular procedure known as trepanation, an operation which will be dealt with in considerable detail in Chapter 9. But what is probably their most impressive surgical achievement is their frequent success in performing the Caesarean section. According to Erwin Acker-knecht, despite many early references to this operation in Western literature, the first authenticated reports of genuine Caesareans (to be distinguished from the practice of removing the foetus from a dead mother) come from the sixteenth century. Although many of the reports of the Caesarean section being undertaken in Africa are little better than rumour or hearsay, some are clearly reliable. Robert Felkin gives an eye-witness report from the nineteenth century:

So far as I know, Uganda is the only country in Central Africa where

121

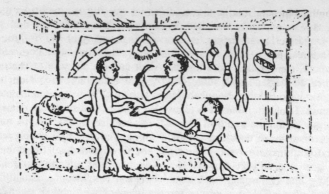

Figure 25

abdominal section is practised with the hope of saving both mother and child. The operation is performed by men, and is sometimes successful; at any rate, one case came under my observation in which both survived . . . It was performed in 1879 at Kahura. The patient was a fine healthy-looking young woman of about twenty years of age. This was her first pregnancy. I was not permitted to examine her, and only entered the hut just as the operation was about to begin. The woman lay upon an inclined bed, the head of which was placed against the side of the hut. She was liberally supplied with banana wine, and was in a state of semi-intoxication. She was perfectly naked. A band of mbugu or bark cloth fastened her thorax to the bed, another band of cloth fastened down her thighs, a man held her ankles. Another man, standing on her right side, steadied her abdomen [see *Figure 25*]. The operator stood, as I entered the hut, on her left side, holding his knife aloft with his right hand, and muttering an incantation. This being done, he washed his hands and the patient's abdomen, first with banana wine and then with water. Then, having uttered a shrill cry, which was taken up by a small crowd assembled outside the hut, he proceeded to make a rapid cut in the middle line, commencing a little above the pubes, and ending just below the umbilicus. The whole abdominal wall and part of the uterine wall were severed by this incision, and the liquor amnii [i.e. amniotic fluid] escaped; a few bleeding-points in the abdominal wall were touched with a red-hot iron by an assistant. The operator next rapidly finished the incision in the uterine wall; his assistant held the abdominal walls apart with both hands, and as soon as the uterine wall was divided he hooked it up also with two fingers. The child was next rapidly removed, and given to another assistant after the cord had been cut, and then the operator, dropping his knife, seized the contracting uterus with both hands and gave it a squeeze or two. He next put his right hand into the uterine cavity through the incision, and with two or three fingers dilated the cervix uteri from within outwards. He then cleared the uterus of clots and the placenta, which had by this time become detached, removing it through the abdominal wound. His assistant endeavoured, but not very successfully, to prevent the escape of the intestines through the wound. The red-hot iron was next used to check some further haemorrhage from the abdominal wound, but I noticed that it was very sparingly applied. All this time the chief 'surgeon' was keeping up firm pressure on the uterus, which he continued to do till it was firmly contracted. No sutures were put into the uterine wall. The assistant who had held the abdominal walls now slipped his hands to each extremity of the wound and secured there. The bands which fastened the woman down were cut, and she was gently turned to the edge of the bed, and then over into the arms of assistants, so that the fluid in the abdominal cavity could drain away on to the floor. She was then replaced in her former position, and the mat having been removed, the edges of the wound : . . were brought into close apposition, seven thin iron spikes, well polished, like acupressure needles, being used for the purpose, and fastened by string made from bark cloth. A paste prepared by chewing two different roots and spitting the pulp into a bowl was then thickly plastered over the wound, a banana leaf warmed over the fire being placed on the top of that, and, finally, a firm bandage of mbugu cloth completed the operation.

Until the pins were placed in position the patient had uttered no cry, and an hour after the operation she appeared to be quite comfortable. Her temperature, as far as I know, never rose above 99.6°F, except on the second night after the operation, when it was 101°F, her pulse being 108.

Erwin Ackerknecht, who gives a shortened account of this operation, despite admitting that it was not an improvised affair but clearly a known procedure in native Uganda, felt somewhat embarrassed at its occurrence among African peoples; it appears to have upset his racial preconceptions and disturbed his simplistic belief in white superiority, and he is almost apologetic that he cannot explain why a black surgeon of the time was able to perform such an operation safely and, by his own admission, in some respects better than many contemporary white surgeons. Alcohol was not the only anaesthetic available to native peoples in Africa or other parts of the world. The Zuñi Indians of New Mexico used the powerful medicinal and psychoactive datura plant as a highly effective means of anaesthetising patients undergoing a variety of surgical operations, including the successful removal of growths from the breast.

On the South Seas atoll of Vaitupu, one of the Ellice Islands (Tuvalu), the indigenous Polynesian people have a remarkable medical tradition that includes a number of surgical operations in its repertoire. Elsewhere in the Pacific, cutting implements were typically made of split bamboo and, in the case of the high volcanic islands, obsidian (a natural volcanic glass with a very sharp cutting edge). Neither kind of material was available to the inhabitants of the coral atolls such as Vaitupu and shell was used instead for many tasks. But the native doctors of the island required a more effective material for fashioning their surgical implements, and for this purpose they selected the teeth of various species of shark. When provided with wooden handles they were used to makc four different kinds of lancets, but their use was, like all other signs of native medicine, forbidden in 1892 when the Ellice Islands became a British Protectorate. Nevertheless, the use of shark's teeth in surgery was carried on among the islanders and the native surgeons continued to be esteemed and would receive patients from distant islands. A number of specific surgical operations are known to have been performed with one of the types of lancet known as the *ponga kiva*. This type of surgical implement was made from a broad serrated shark tooth. Such lancets were particularly useful for making long incisions, when they were tapped with a mallet during cutting procedures. The *ponga kiva* was used when removing leprous tissue and in operating on swellings caused by the contagious disease yaws, tubercular glands in the neck, subcutaneous lipoma (fatty tumours) and elephantoid scrotum. Some of the procedures involved in such operations are described by the anthropologist Donald Kennedy:

The removal of a subcutaneous lipoma was accomplished in this manner. An assistant to the operating surgeon was provided with a pair of smooth-edged rectangular strips of turtle's carapace (*fau*). Holding one of these in each hand on either side of the lipoma, he pressed downward and inward so that the tumour was pinched up and raised well above the surrounding parts. This also served to distend the skin over the tumour and thus make the action of the lancet easier.

The incision was made straight across the tumour so that the skin and surface-tissues fell apart, exposing the core. Blood was removed by the assistant who held the *fau*. This person, bending from time to time over the wound, would blow out the blood which threatened to obscure the course of the lancet. When the whole extent of the lipoma became visible, a fish-hook [actually a specific surgical instrument originally made of wood and based on the shape of a local fish-hook] attached to a line was inserted in it and the line was given to a third assistant who held it taut, thus drawing up the tumour while the *tufunga*, or master surgeon, undercut it with his *ponga kiva* until it was quite free. The wound was then covered with soft strips of beaten coconut-husk which were bound in place by strips of pandanus-leaf, forming a rude bandage . . . abnormal swelling of tubercular glands in the neck, was operated upon in much the same manner.

In the removal of elephantoid scrotum the *fau* were not employed. The testicles, having been pushed back with the finger tips, the growth was removed with the *ponga kiva* lancet, the incision being made to encircle the tissue to be removed. The subsequent treatment of the wound was the same as that described above. It is interesting to note that my informant insisted that, to his positive knowledge, death never ensued. The mortality following this operation under modern operating-theatre conditions is said to be no less than 5 per cent.

Kennedy's statement, written in the late 1920s, on the higher success rate of the Polynesian surgeon compared with his contemporary Western counterpart in the performing of the last of these operations echoes the skill of the Ugandan surgeon in performing the Caesarean section. Clearly, tribal surgeons were able to perform at least some operations, without all the finery of Western surgical equipment, equally if not better than the 'civilised' surgeons of the time. In Kennedy's opinion the anatomical knowledge of the surgeons and also that of the specialist masseurs of Vaitupu could not have been obtained without post-mortem examinations although the practice of performing autopsies had apparently died out. As is also the case among the Aleutian Islanders, massage was an integral part of the indigenous medical practices (particularly in midwifery) on Vaitupu. It consisted of a number of techniques, such as pinching, pressing, kneading, etc., applied by both the hands and feet of the masseur, and was reported to be highly therapeutic.

Chapter 9

Stone Age Surgery

Undoubtedly the widest-known major surgical operation in tribal cultures is trepanation (also called trephination or trephinement), which, as will become clear, was also known in the Stone Age. This operation involves the removing of one or more parts of the skull without damaging the blood vessels, the three membranes that envelop the brain – the dura mater, pia mater and arachnoid – or the actual brain; not surprisingly, it is a procedure that requires both skill and care on the part of the surgeon. The word trepanation is derived from the Greek *trypanon* meaning 'borer' or 'drill', a reference to the trepan or trephine, the surgical instrument used in one of the methods of performing the operation. Thomas Wilson Parry MD (1866–1945) was a doctor who was particularly fascinated by trepanation, especially in the way it was performed in the Stone Age (see Plate X), and he published numerous accounts of his researches into this curious practice and its prehistoric origins. In 1918 he even published a humorous ballad concerning Neolithic trepanation, which, whilst somewhat lacking in lyrical accomplishment, nevertheless manages to evoke some details of the operation as it may have been performed (as Parry thought, to expel a demon that was believed to be the cause of epilepsy). The following verses extracted from this ballad contain the essential surgical features, beginning with the announcement of the Neolithic medicine-man:

> 'This patient must be *now* trephined,
> Let all the others go;
> To-morrow when the sun is up
> My magic I'll them show.'
>
> Two men the epileptic bore
> And laid him on a trunk,
> And when the wretch was coming round
> He showed some signs of funk.
>
> No questions put they to the man;
> The doctor cleared his throat,
> Then, bringing flints from out his hut,
> Took off his hairy coat.
>
> A crowd had gathered all around,

To watch the bloody deed;
Their curiosity was stirred
To see his devil freed.

With sharp flint flake the surgeon made
A cruciform incision;
The blood did spurt, the wound it hurt,
The crowd laughed in derision.

The two assistants pressed the flaps
To stop the blood from running;
The Medicine-Man did scheme and plan,
He was so full of cunning.

He scraped the pericranium,
Until the skull was bare;
Then scratched the bone with a sharp stone,
It did not matter where.

He scraped that bone and scratched and scraped –
The scratches made a groove,
The groove a basin-like ellipse.
The patient did not move.

The fact was this, when he came round
So rotten did he feel,
He fainted when he found himself
The centre of such zeal.

The hollow soon became a hole,
'Twas all but through the bone,
His diploë,* you well might see,
But still he made no moan.

The inner table only now
Protected his soft brain,
One final scrape and he did make
That hole a window-pane.

The devil stirred within his skull
And, with a fearful yell,
Escaped from out its prison-house
To seek its own in hell.

On a more serious note, Parry not only identified the various methods by which trepanation was performed but also conducted his own experiments on skulls using a variety of surgical implements of a traditional kind in order

* The diploe is the soft layer of bone tissue between the hard inner and outer layers of the bones of the skull.

to judge for himself the efficiency of each. Needless to say, the experimental trepanations he undertook were done on skulls not of the living but of the dead. Parry conducted about fifty experiments on both recent and older (and hence drier) human skulls, using implements made of obsidian, flint, slate, glass, shell and shark teeth (some of which are shown in *Plate XI*). He identified four distinct methods of trepanation.

The first method, which seems to have been the most common in Neolithic times, consisted of scraping a hole in the skull by means of a scraping tool in the form of a small flake of obsidian, flint or another material held between the thumb and forefinger. The procedure started with the making of a V- or Y-shaped incision, which was then made into a groove by repeating the action. Next, the edges of the groove were worked on with a curved movement of the scraper which eventually produced an elliptical-shaped depression (as mentioned in the verses cited above). This depression was scraped until it was circular in shape, and when the inner table was reached, the roundel of bone could then be removed from the skull. Parry records that the average time it took him to perform a trepanation by the scraping method on a fresh adult skull was half an hour. He found both obsidian and flint excellent materials to work with surgically, and also expressed the opinion that shells – which were used in Oceania to perform such operations – were highly effective too, noting that he trepanned the skull of a nine-month-old infant in twenty-five minutes with a beach-worn oyster shell and that of an adult Maori in a mere thirteen and a half minutes with a larger and more robust shell.

The second method of trepanation involves a mixture of boring and sawing. With the trepan a number of closely knit holes were bored in the form of a circle, and then the ridges of bone between the holes were cut or sawed to complete the circle, thus facilitating the removal of the roundel. The use of this particular technique by Stone Age surgeons can be readily deduced from the ragged shape and uneven surface of the circular edge (caused by the alternation between the bored holes and the ridges between them along the circumference) left on the skull after the roundel has been removed. He found this boring and sawing procedure to be much slower than the scraping method, although it did have the advantage of permitting the making of larger holes than by the first operating technique. The tedious and time-consuming boring of the circle of holes by hand is made much easier and quicker if a bow-drill is used (for a description of the bow-drill and its use, see Chapter 11). As the remains of Neolithic bow-drills have been found (and are preceded by such drills in the Maglemosian culture of the Mesolithic period), it is clear that Stone Age peoples made use of this mechanical contrivance in various practical ways and therefore could easily have used them in some surgical operations, including trepanations. Parry found that the most effective of all surgical implements for boring holes in bone, with the sole exception of those made from metal, was a hafted shark's tooth. He favoured the shark's tooth due to the sharpness and strength of its point and its serrated edges, but also because it has a natural flange. This last feature is particularly useful in its role as a trepan, since it can be used in such a way that the point does not go too deeply into the skull and thus risk irreparable damage to the membranes or brain beneath.

The third procedure described by Parry is called the push-plough method, which he believes was done by using a beaked flint instrument which was pushed repeatedly along the surface of the skull in an oval or circular path. What was initially a line would become a groove, and then a furrow, the more the flint was used. When the furrow had penetrated to the required depth, the piece of bone to be removed could then be prised out. This method is said to be especially good for the removal of large roundels from human skulls. The fourth prehistoric method, used in ancient Peru, was the sawing out of a rectangular piece of the skull by four straight cuts. This is a particularly difficult and therefore dangerous method, since the slightest error could endanger the patient's life. According to Parry, there is no evidence that this method was used at all in Neolithic and later prehistoric Europe.

Having described how trepanning operations were performed, it remains to describe why such procedures were undertaken. There are, in fact, numerous reasons why this operation has been performed in various parts of the world and at different times. It is known from historical and ethnographic sources that trepanation has been undertaken to treat epilepsy, mental illness, demonic possession, fractures, severe and recurrent headaches, vertigo and deafness, as well as for the removal of foreign bodies and even as a supposed aid to prolonging longevity. The reasons why it was performed in prehistoric times remain necessarily obscure, but one, and probably more, of the reasons listed above is likely to have been involved in Stone Age trepanning. The reader will no doubt agree that the operation sounds extremely painful, but surprisingly those who have studied, observed and even undergone trepanation do not concur in this respect. The shock involved in the operation is said to be fairly insignificant. Parry mentions a Polish doctor who carried out experimental trepanation on both cats and dogs in order to assess the level of shock induced in the animals, which turned out to be very little. Hilton-Simpson, who conducted research into the practice of trepanation among the Algerians, personally witnessed an operation on a small girl, 'who, becoming restive and noisy, was only silenced into operative subjection by a terrible outpouring of oaths and foul language administered by the operating surgeon himself'. The Algerians are elsewhere reported not to use any antiseptic or anaesthetic agents while trepanning their patients. Another scholar who has investigated the practice thinks that anaesthetic would have been unnecessary since most patients enduring trepanation would have already been in a coma. This seems, in the light of both the case of the Algerian girl and the East African trepanations cited below, to be incorrect.

Trepanation is still regularly practised among the Gusii of Kenya, a Bantu people with a population of about one million, and theirs is perhaps the last surviving traditional practice of its kind. The operation is performed primarily to treat accidental injuries today (particularly in remote areas far from the nearest hospital), although previously many wounds received during violent conflicts with neighbouring peoples were also treated in this way. Rolf Meschig, who in recent years has made a first-hand study of Gusii trepanation, discovered that one of the native surgeons – or 'skull-openers' as they are known – had been performing about 50 such operations a year. Meschig witnessed a trepanation which took eleven hours, during which time the patient was fully conscious and able to endure whatever pain was involved without recourse to any anaesthetic. Herbal medicines are used by the Gusii surgeons both to control bleeding and to clean the wound. Since the majority of trepanations are successful it seems that some of the plants used must be effective antiseptic agents. For the need for an antiseptic is even more important than for an anaesthetic. Earlier research into Gusii surgery suggests that the *omobari omotwe*, or head surgeon, was, like many doctors in our own societies, well paid for his professional services. One patient is reported not only to have paid a considerable sum of money, but also given the surgeon a sheep, a goat, three chickens and three four-gallon drums of local beer.

The most remarkable case of trepanation among the Gusii is that of a local policeman who had struck his head accidentally on a door lintel around 1940. For the next few years he suffered from severe headaches, and in 1945 had the first of a number of trepanations to relieve his condition. The result of these repeated operations can be seen in *Plate XII*; the missing area of the skull is some 30 square inches. This unfortunate man reported experiencing considerable pain during his various operations, in contrast to the operation witnessed by Rolf Meschig, which seems to have been comparatively painless.

Clearly whether or not a particular trepanation is painful relies on a number of factors: the general health of the patient, the skill of the surgeon, the size of the trepanned area and so on. The success rate of the operation among the Gusii is indicated by the actual *omobari omotwe* who operated on the policeman. He had been taught the scraping technique by his father (and expressed his disapproval of the sawing technique) and had performed it successfully more than a hundred times and said that he had never lost a patient. In his experienced eyes the only reason for trepanation was chronic headaches precipitated by a blow to the head. Multiple trepanations are also reported from the South Pacific. An individual in New Britain, a large island off the coast of New Guinea, is recorded as having eight separate holes in his skull. In the Loyalty Islands, plates made from coconut shell are used to cover such holes; one man is said to have five such plates on his head.

In the Americas trepanation was the speciality of the pre-Inca and Inca inhabitants of ancient Peru. It was performed by a scraping tool called *tumi*, which was made of obsidian or in some cases of metal. The examination of a pre-Columbian skull from Patallacta in the Peruvian Highlands with five trepanned holes in it showed that only one of these had any signs of possible infection. This would seem to indicate that at least the first four trepanations were successful, and if this is so then some effective antiseptic must have been known to the ancient Peruvian surgeons. Furthermore, bearing in mind the fundamental spiritual and practical role of the coca plant in their culture, it is also likely that the Peruvians made use of its cocaine-bearing leaves as a surgical anaesthetic. In fact, systematic analyses of large numbers of ancient Peruvian trepanned skulls have revealed a remarkably high success rate for this difficult operation. An examination of 214 such skulls reported by T.D. Stewart indicated that 55.6 per cent had healed completely after trepanation, 16.4 per cent were in the initial stages of healing (when the individual died – not by any means necessarily due to having been operated on) and 28 per cent showed no signs of healing. A similar study of 400 skulls by Rytel found that 62.5 per cent (250) showed signs of having healed.

To give some idea of how this compares with the success rate of the same surgical procedure in Europe, it should be said that trepanation even at the beginning of the nineteenth century was the cause of some trepidation, as the success rate was so low. It was only with the introduction of modern anaesthesia, antisepsis and asepsis, around the middle of the century, that the mortality rate for those undergoing trepanation fell from 43 to 14 per cent. F. P. Lisowski is even more pessimistic about the efficacy of such European trepanations, stating that: 'For a long time medical science doubted the existence of healed prehistoric trepanations, since eighteenth and nineteenth century surgeons of the pre-antiseptic era rejected this procedure owing to the almost one hundred per cent mortality rate.'

Thus ancient Peruvian surgery – at least in regard to the practice of trepanation – was more efficient than that of the highly 'civilised' surgery of the early nineteenth century. Studies of both the trepanned skulls of the Incas and some of those found in Neolithic Europe indicate that healing seems to have been the norm in both cases. It appears that the trepanations performed in Europe during Neolithic times sometimes had the same chance of success as those in the earlier part of the nineteenth century. Now that's progress! It is hard to explain the Stone Age success rate without concluding that some kind of effective antiseptic agent must have been used. Furthermore, the surgeons of the time must have understood the need for it. In 1940 the prehistoric archaeologist Stuart Piggott, whilst acknowledging that trepanation was both a dangerous and a difficult operation, believed that such surgery in the Neolithic was a rather haphazard affair and that it was sometimes 'accidentally successful'. Yet it must now be acknowledged that a greater competency in trepanation than Piggott would allow must have been behind the success rate of the operation in the Neolithic.

As I have already said, when the first prehistoric trepanned skulls came to light, it was not thought that they could be evidence for cranial surgery, and so various more or less plausible alternative explanations were sought in order to account for the making and function of the holes. A general practitioner named Prunières discovered his first Neolithic trepanned skull in the Lozère, France, in 1865. It was clear to him that the hole in this skull had been deliberately made, as its edges were smooth and appeared to be polished. He surmised that it could be a drinking cup and that the hole was the aperture that had been made to be raised to the lips. Other trepanned skulls were interpreted as wounds caused by swords and other weapons, or the result of accidental trauma. In the mid-1870s the highly esteemed French surgeon and physical anthropologist Paul Broca realised the true nature of both such holes in various Neolithic skulls and the roundels of bone that were removed from them. Yet more unusual-shaped openings in the skull continued, on occasion, to perplex archaeologists. In 1960 Kathleen Kenyon discovered some skulls in Palestine dating from the sixth century BC which had small quadrilateral sections removed. Her initial reaction was that they may have been made during the course of some terrible experiments carried out on captives, but she subsequently identified them as the result of trepanning war wounds.

It was also once believed that the practice of trepanation may have begun in the Neolithic period in the realm of veterinary surgery and was then subsequently adapted for use on human patients. The operation has been performed on domesticated sheep into modern times, mainly to treat the disorder known as staggers, in which the animal staggers around due to the presence of the larvae of tapeworms under the scalp. In Romania, for example, shepherds trepanned their sheep to remove the larvae that were afflicting them, apparently with little success in many cases where the larvae had already entered the animal's brain. Lisowski believes that such procedures performed on human subjects may be the origin of diverse strands of folklore as far afield as the folk stories of the Balkans that describe the removal of beetles from the human head by means of trepanation, and the extraction of centipedes from the head that is mentioned in early Tibetan medical texts.

There is also an early Chinese source which mentions that the Ta-Chhin (i.e. the inhabitants of Roman Syria) are able to open the 'brain' in order to extract worms that are said to be the cause of blindness. The leading authority on the history of Chinese science, Joseph Needham, cited this as the sole instance of any interest in Western medical practices by the Chinese, but other early references may indicate that trepanation was practised by the Chinese themselves in ancient times. Although humans have contracted numerous disorders from their domestic livestock from the Neolithic to the present (recent controversy over BSE or 'mad-cow disease' in Britain and elsewhere is hardly likely to be the last such dangerous by-product of animal domestication), human trepanation is too widespread a practice to be explained as originating with veterinary interventions of the same kind, and such a solution to the origin of trepanation could only ever be partial. In Melanesia and other parts of Oceania, trepanning is practised among peoples who never tended sheep, goats or other ruminants that are susceptible to staggers, and thus could hardly be said to have originated in the veterinary sphere.

The earliest reported trepanned skull is that belonging to a skeleton discovered under a tumulus at Colombres in Spain, dated to the Mesolithic period, about 8000 BC, and identified as belonging to the Asturian culture of the region. An oval hole the size of a small coin indicates the possibility of trepanation, although it is not clear whether this was made before or after the death of the individual. If it were the latter, then this would not qualify as a bona-fide case of trepanation. If it was made during life, then the operation was a failure as there are no signs of healing to be found on the skull. To date the earliest clear-cut case of trepanation was discovered only very recently, in September 1996, at the Neolithic burial site of Ensisheim in Alsace, France, and dated somewhere between 5100 and 4900 BC. Among the various human remains found at the site were those of a 50-year-old man (known as Burial 44) whose skull had two holes in it, both of which were clearly the result of trepanning. Both displayed clear evidence of healing, and Kurt Alt and his colleagues, who have researched the skull, consider it to be the oldest healed neurosurgical operation known in the world. They remark that 'its technical realisation testifies to the high craftsmanship and well-founded anatomical knowledge of the surgeon'. There are hundreds if not thousands of other trepanned skulls from all across prehistoric Europe (as far afield as England, Sweden, Italy and Russia), which shows that the practice was well known and not just the result of sporadic innovations by a few individuals. Other instances of prehistoric trepanned skulls with multiple holes have been found, and one Neolithic skull from Germany has a large part of the top of the skull surgically removed, rather like the Gusii policeman mentioned earlier in the chapter.

Curiously, after the Neolithic period the practice of trepanation goes into decline in Europe, and although there are Bronze Age, Iron Age, Roman, Viking, Merovingian and medieval examples of the operation, they are comparatively few and far between. There is likewise a continuous although rather insignificant trail of trepanned skulls and reports of 'cutting out of the fool's stone', as the operation was described in the seventeenth century, that leads us into the modern era that I have already mentioned. We can see from this that the great flourishing of the trepanning tradition before the modern era of surgery can be clearly traced to Neolithic times. Was it an innovation of this period, or did it develop out of some Mesolithic or even Upper Palaeolithic surgical knowledge? It also needs to be asked whether trepanation as a surgical procedure was invented independently in the various areas of the world where it has been practised extensively (i.e. pre-Columbian Peru, Melanesia and Neolithic Europe), or did it arise first in only one place, from which it subsequently spread? In short, is the early emergence of trepanation due to independent invention in the three areas or diffusion from some common centre?

In his discussion of such questions in a 1940 paper, Stuart Piggott came to the conclusion that although the curious practice of trepanation seemed an improbable candidate for independent invention in more than one area (the idea being that it was strange enough that it had been invented even once), this seemed the least unlikely alternative available. Don Brothwell of the University of York, England, has recently reviewed this question and sees three possible explanations for the distribution of the practice of trepanation. Firstly, as believed by Piggott, it could be due to independent invention, though Brothwell sees this as highly unlikely. Second, it may originate in the Palaeolithic era, which would imply that the Palaeo-Indians who crossed over the Bering land bridge as the first people to enter the Americas brought some surgical knowledge of the practice with them, thus providing an initial tradition of New World trepanation which the ancient Peruvian surgeons would have drawn upon. It is not only the complete lack of trepanned Palaeolithic skulls that makes Brothwell reject this option. He also points out that if the practice were indeed a Palaeo-Indian trait, then why are trepanned skulls not really found outside the so-called advanced societies of Peru and to a lesser extent Mexico?

Brothwell believes that a third possibility can better explain the extremely limited distribution of trepanation in the New World. He suggests that the trepanning traditions of the New World can be explained by early historical diffusion either across the Pacific or across the Atlantic. This theory owes a debt to the heretical theories and dramatic instances of experimental archaeology of Thor Heyerdahl. Heyerdahl thought that the navigational and boat-building skills of the ancient world were more than adequate to make ocean crossings, and he made boats based on ancient designs and set sail to prove his point. He crossed the Atlantic in the reed-boat *Ra* II and the Pacific on the balsa raft *Kon-tiki*. The successful completion of these extraordinary voyages upset a great many armchair theorists who had thought such things were impossible. No doubt the huge attention of the world's media and the string of best-sellers that Heyerdahl produced (not to mention the winning of an Oscar for his film of the *Ra* expedition) also rankled among many in the academic community. Although the critics were silenced in respect of the plausibility of such voyages, few people believe that there is any substantial evidence that such ocean crossings were actually undertaken in ancient times; the possibility was one thing, proving it happened was another kettle of fish altogether. So how does this fit with the distribution of trepanation? Not particularly well, as it turns out.

Firstly, if we look at the possibility of the introduction of the practice into the New World from across the Pacific, it must be asked who could have introduced it. Although suggestions have been made that ancient Chinese seafarers made contact with the Americas, this is by no means a generally accepted fact, and anyway such theories are mainly limited to proposing contacts with native Americans of the North Pacific and not with the cultures much further to the south. The evidence for the existence of trepanation in early China is also sketchy at best, so this seems a very unlikely scenario. The only other area in the Pacific Rim that has a trepanning tradition of any real significance is Melanesia, and no one has, to my knowledge, suggested that they ever initiated contact with the Americas. Furthermore, Melanesian trepanation is known from ethnographic rather than archaeological reports and so we cannot even be certain that it was practised in the prehistoric cultures of the region, although it would not be surprising if this turned out to be the case. Yet clearly the idea that trepanation was introduced to the New World from across the Pacific in pre-Columbian times is without any firm foundation.

The other possibility, namely the theory of the Atlantic crossing, is equally untenable. Although there are reports of one or two trepanned skulls from ancient Egypt, it is clear that this surgical operation was of little or no significance in ancient Egyptian medicine. So even if the ancient Egyptians ever reached the New World, it is hardly likely that they would have introduced a surgical practice that they barely – if at all – knew themselves! Whether the Phoenicians or other seafaring peoples who have been put forward as candidates for trans-Atlantic crossings ever showed much interest in trepanation is also highly dubious. The theory that trepanation was imported to the New World during early historical times has no evidence to support it and so this third of Brothwell's three explanations for the distribution of the practice can be firmly rejected, leaving only the first two options.

Until very recently it was believed that trepanation was not practised in Aboriginal Australia, but the recent discovery of prehistoric skulls with holes made by this means has shown that the practice did indeed exist on this continent. The exact dates of these trepanned skulls are not known, but it is thought that they are no more than a few thousand years old. Although it may turn out that the Australians were introduced to trepanation by the Melanesians, there is no evidence for this, and so, with the present state of knowledge, it seems to support the idea for independent invention rather than diffusion in Palaeolithic times. Nevertheless, the Australian finds show that hunter-gatherers also practised trepanation and that the operation need not have waited until the advent of farming for it to have been undertaken in prehistoric Europe and elsewhere. Yet whichever of these two options for the widespread practice of trepanation turns out to be the correct one, the surgical skill that it involved can be clearly shown to exist in the prehistoric periods of diverse parts of the globe. We can only agree with Thomas Parry's eulogy to this unusual example of prehistoric innovation:

I think we owe much to all daring pioneers in whatever sphere of action they may endeavour to excel ... So: – To the memory of these bold empirics who, in the dawn of the world, dared to *do*, and who were the first to lay the foundation of our present-day cranial surgery, I raise the glass to my lips, in admiration of their pluck and endurance, in a silent but reverential toast.

As remarkable a surgical operation as trepanation is, its unique place in the evidence for Stone Age surgery is probably greatly exaggerated. Most of the surgical procedures and ritual alterations of the body that are known to have been among the repertoire of various tribal peoples would also have been well within the capacities of prehistoric peoples (as proved by the Stone Age evidence for trepanation), but would leave little evidence behind for the archaeologist and palaeopathologist to work on. As it is usually only the bones of Stone Age people that survive to be discovered by archaeologists, any operation that was performed on the soft parts of the body only cannot be detected. It seems likely that various operations were performed at least from the Neolithic period onwards, and perhaps earlier; to believe that they restricted themselves only to the comparatively difficult trepanning of skulls seems extraordinarily unlikely.

Another kind of medical practice that we do know was performed in the Neolithic period in Europe is dentistry. A visit to the Neolithic dentist may have inspired a similar degree of fear and trepidation as it does even today. We tend to think of early historical dentistry as being performed in a very painful fashion. In caricature the early dentist is often depicted as having assistants whose job is to tie down or otherwise restrain the hapless patient, whose tooth is removed – without anaesthetic – by an inordinately large pair of pliers, a crude picture that is coloured in with copious amounts of blood. Yet as we shall see, various anaesthetics were available even in Neolithic times and so the realities of prehistoric dentistry were probably not so horrendous as our imagination may lead us to believe. Whilst there is no evidence of dentistry prior to the Neolithic, there is evidence that people in the Mesolithic period were concerned with looking after their teeth. Analysis of some Danish skeletons dating from this time has revealed the presence of grooves between the molars caused by the use of toothpicks. If we are to believe that chewing gum may benefit the health, then the prehistoric gum made from the inner bark of the birch tree discovered at the Mesolithic site of Star Carr in Yorkshire, England, may also be considered as an aspect of Stone Age hygiene. Neolithic chewing gum, with teeth marks on it, has been found in Switzerland.

Whilst conducting palaeopathological studies on Danish skeletons from a number of prehistoric and early historic contexts, Pia Bennike was surprised to discover a round conical hole in one of the upper molars still *in situ* in the jaw of a Neolithic skull belonging to a man and dug up in 1960 at the passage-grave site of Hulbjerg, on Langeland in Denmark. The remains of more than 50 people were found at the site and although this particular skull was the only one to show signs of a dental operation, one case of trepanation was also found, showing that different types of operations were performed in this Neolithic community. It seemed to Bennike that the hole in the molar tooth was made by a bow-drill, and we may therefore see this as the prehistoric prototype of the dreaded dentist's drill of our own era. Meticulous examination revealed that the probable reason for the performing of the dental treatment was the infection of a deep cavity in the tooth which had probably resulted in a painful abscess. In an illuminating and unusual act of experimental archaeology designed to establish beyond reasonable doubt that the hole in the tooth was indeed made by such a method, a replica bow-drill was made and with it a hole was bored in a recently extracted tooth. A comparative examination of the Neolithic hole (see *Plate XIII*) and that made in the modern tooth under a scanning electron microscope (SEM) showed that they were very similar; in fact, strikingly so. It was also clear that the hole in the Neolithic tooth had been made during its owner's lifetime and not after death, as a small amount of calculus was found on it, something which would not have been possible if the man had been dead when the hole was bored. Whilst the bow-drill was used in Neolithic dentistry it had a variety of other uses in boring holes in materials, and was also one of the early means of engendering fire, discussed in detail in the next chapter.

Although some of the individuals who have studied trepanation have said that it was comparatively painless, the Gusii policeman (who, with his numerous operations, must surely qualify as an expert in these matters!) clearly suffered considerable pain. Dental operations are, as we all know, painful and if prehistoric trepanation did not always require an anaesthetic then drilling teeth probably did. There is clear evidence that plants with anaesthetic properties were widely used in ancient times. Alcohol was certainly used as such, often in conjunction with other psychoactive substances. Dioscorides, writing in the first century AD, mentions wine mixed with extracts of the mandrake plant (*Mandragora*) as being the standard surgical anaesthetic of his day. In ancient Egyptian mythology there is an incident in which the god Ra overcomes the goddess Hathor by stupefying her with mandrake beer. Beer was brewed by both the Predynastic Egyptians and by the early Sumerians, and both beer and wine have their origins in the Neolithic period, extending back to the fourth millennium BC and perhaps even earlier.

During the period from about 3500 to 3000 BC, the Bronze Age cultures of the eastern Mediterranean area were consuming wine from metal vessels. Their neighbours to the north (who were still following a Neolithic way of life) were converting to the mixed blessings of alcoholic beverages and imitated the shape of these metal cups in their own ceramic vessel designs. Alcohol use spread across Neolithic Europe, gradually displacing the use of other psychoactive substances in its wake. It appears that as it was introduced to the various parts of the continent, it was initially used in conjunction with mind-altering plants such as the opium poppy (*Papaver somniferum*) and cannabis (*Cannabis sativa*). As the new intoxicant took hold, the use of these other substances declined. For although the drinking of alcohol was a Stone Age innovation, it was, nevertheless, a later phenomenon than the use of opium.

The opium poppy, the source of both morphine and heroin, seems to have been domesticated by Old European farmers in the western Mediterranean area from about 6000 BC. That the cultivation of the opium poppy spread westwards during the Neolithic period is indicated by numerous later finds of its seeds from Switzerland, Germany and elsewhere. By the Iron Age it was also present in more northerly regions such as the British Isles and Poland. The seeds of the opium poppy may well have been used in baking and their oil been pressed into use for cooking or lighting during prehistoric times. Yet these are clearly minor uses of the plant, and the Stone Age interest in it must have been for its psychoactive and medicinal properties. In many non-Western cultures, magic and medicine are often two sides of the same coin, and in prehistoric times opium was probably used to relieve pain as well as to enter into altered states of consciousness for spiritual insight. Opium appears to have played a significant role in the religious life of Old Europe. The Cueva de los Murciélagos is a Neolithic site at Albuñol, Granada, in southern Spain, dated to about 4200 BC. Inside woven grass bags found with the burials were a large number of opium poppy capsules, and this discovery suggests that opium may have been an active part of funerary rituals. Certainly the placing of the capsules with the bodies points to a clear association between opium, altered states and the realm of death. This indicates that the use of opium in the ancient world (for example in the rituals of Minoan Crete) may have been an outgrowth of an Old European practice.

The use of cannabis or hemp can also be traced back to the Stone Age. The cannabis plant is native to Central Asia but had already spread across the Old World before history began. As well as having psychoactive properties the cannabis plant also provides an extremely strong fibre, which has been used from time immemorial. Nevertheless its mind-altering effects were also made use of in Neolithic times. Stone Age cultures on the steppes used it in a ritual fashion at least as far back as the third millennium BC. In a burial site in Romania belonging to the Kurgan people (identified by Gimbutas as the Proto-Indo-Europeans), archaeologists discovered a small ritual brazier which still contained the remains of charred hemp seeds. This, like the use of opium in Old Europe, seems to be a practice that is ancestral to those known from historical sources. In the fifth century BC, the Greek historian Herodotus gives the following report concerning the Scythian nomads north of the Black Sea:

> On a framework of tree sticks, meeting at the top, they stretch pieces of woollen cloth. Inside this tent they put a dish with hot stones on it. Then they take some hemp seed, creep into the tent, and throw the seed on the hot stones. At once it begins to smoke, giving off a vapour unsurpassed

by any vapour bath one could find in Greece. The Scythians enjoy it so much they howl with pleasure.

Herodotus has often been seen as an unreliable source, but in this particular instance at least, he has been shown to have been accurate, in all but one respect. At the other end of the Scythian world Soviet archaeologists discovered archaeological evidence to support his account. The excavation of Scythian tombs at Pazyryk in the Altai mountains of southern Siberia (dating from the fifth century BC) revealed metal braziers, the burnt remains of cannabis seeds and even the poles used to support the tent! This not only proved Herodotus right, but also showed how wide an area this cannabis cult was practised in. The only detail Herodotus got wrong was to say that they threw hemp seeds on the brazier. This would not have had the desired effect as the seed is the only part of the cannabis plant that is not psychoactive. The presence of charred seeds in both the Kurgan burial and the Scythian tomb indicates that the combustible (and psychoactive) parts of the plant – namely flowers and leaves – had been consumed and the hard residue left behind.

Cannabis not only went west to Europe from its homeland on the steppes but also travelled to China. Linguistic research undertaken by the Chinese scholar Hui-Lin Li indicates that both the technological and the psychoactive uses of the plant were known to the ancient Chinese. In Chinese, hemp is referred to as *ta-ma*, meaning 'great fibre' (*ma* = fibre). Li also points out that in archaic times the character *ma* had two meanings. The first of these was 'chaotic or numerous', a reference to the appearance and quantity of its fibres. The other meaning was 'numbness or senselessness', a reference to its stupefying qualities, which were apparently made use of for medicinal and ritual reasons. The current state of knowledge concerning the prehistoric use of cannabis indicates that it was first cultivated in north-east Asia both for its fibre and also as a means to induce ecstasy among shamans. There are a number of references in ancient Chinese writings to the use of cannabis by magicians and Taoists, and it appears that such uses stem from their shamanistic forebears.

In south-east Asia the earliest known use of a psychoactive substance concerns the practice of betel-chewing. This stimulant is estimated to be taken by 10 per cent of the world's population. It is particularly popular in India, mainland south-east Asia, Indonesia and Melanesia, and is usually taken in the form of a quid (similar to a 'chew' of tobacco). The basic preparation consists of a leaf of the betel plant (*Piper betle*) in which the seed of the areca palm (*Areca catechu*) is wrapped. In order to release the stimulating properties contained within the preparation, an alkaline additive such as slaked lime is mixed with the areca seed. Many users have a quid in their mouths almost constantly, and heavy and habitual use causes the teeth to turn black. Traditionally in the Philippines having black teeth (i.e. being a heavy user of betel mixtures) was a sign of social status. The earliest archaeological evidence for the practice comes from the Spirit Cave site in north-west Thailand where *Piper* seeds were found at levels dated to between 5500 and 7000 BC. Direct evidence for the actual use of a betel mixture comes from Duyong Cave on Palawan in the Philippines. In this cave the skeleton of a man (dating to 2680 BC) was found interred along with half a dozen bivalve shells containing lime, and his teeth were stained as those of any serious betel user should be.

Other psychoactive substances that have been prepared with alkaline additives include coca, tobacco and pituri. In South America the coca plant (*Erythroxylum coca*), the source of cocaine, has been used for thousands of years for its stimulating and anaesthetic properties and may well have been used in trepanation operations in ancient Peru. The Australian Aborigines made use of various native species of tobacco and another nicotine-bearing plant they call pituri (*Duboisia hopwoodii*). Pituri was highly valued in traditional times and was traded over vast distances and across many tribal territories. The Aborigines used *Acacia salicina* as their alkali source to release the nicotine in the pituri plant. The fact that *Acacia* has an extremely high alkali content indicates that a great deal of experimentation with sources of alkali was undertaken before the best available source was identified. There are also other indications that the Aborigines had a detailed knowledge of the chemistry of pituri, for they developed a complex process of harvesting and curing it in order to maximise its psychoactive properties.

The Aborigines never took up the practice of agriculture before the arrival of the white man on their continent. Yet the fact that they paid an inordinate amount of attention to pituri has implications for the origins of agriculture. That what can be seen as a first footstep towards agriculture in Australia involves not a food plant but a psychoactive one is highly significant. The standard theory concerning the origins of agriculture is that this change of lifestyle was primarily concerned with food production. The Australian evidence may lead us to think otherwise. The Oxford archaeologist Andrew Sherratt has suggested a similar genesis of agriculture for the Near East and notes with particular reference to Neolithic Jericho that the first cultivated plants were not perhaps cereals at all but more valuable and portable commodities. He suggests a number of narcotic plants like mandrake, henbane and belladonna (deadly nightshade) as possible candidates.

There is evidence from the New World to support the idea that, at least in some parts of the world, the first plants to be domesticated were not staple foodstuffs but psychoactive species. Many native North American peoples such as the Blackfoot traditionally disdained agriculture and only made an exception in the case of tobacco. This pattern is also corroborated by the prehistoric origins of tobacco use. The native habitat of tobacco is in the lowlands of Patagonia, the Pampas and Gran Chaco; that is to say, the southern part of South America. It was in this region that tobacco was first cultivated by Indians in their gardens some 8,000 years ago. It seems also to have been the case that in this area horticultural practices were first initiated in order to ensure a steady supply of tobacco rather than foodstuffs.

Although the advent of horticulture and agriculture seems to have been brought about in part by the desire for psychoactive substances, the use of mind-altering plants no doubt goes back to primeval times. With the possible exception of the use of the stimulating plant *Ephedra* by Neanderthals (for which see Chapter 16), there is no concrete evidence for the use of psychoactive plants in the Palaeolithic period. Some researchers have claimed that some of the images in the Upper Palaeolithic cave art in France and Spain were inspired by hallucinatory experiences, but this is difficult, if not impossible, to prove. No clear depictions of psychoactive plants or fungi appear in the art of Upper Palaeolithic times and no palaeobotanical remains of such species have been found in archaeological sites dating to this period. The highly mobile hunter-gatherer societies of the Upper Palaeolithic period naturally did not leave such clear evidence of their use of plants as the later Neolithic farmers who lived in permanent villages. No doubt the refinement of palaeobotanical techniques will soon produce evidence for the use of mind-altering plants in the Upper Palaeolithic. The ultimate origins of the use of such substances remain obscure, but Louis Liebenberg has suggested that the practice of smoking could perhaps be traced back to the domestication of fire by early man. Many tribal peoples traditionally transported fire in a small container such as a shell or a wooden tube. Perhaps the burning of a fragrant grass or other plant in such a container led to the development of the smoking pipe in some remote period of prehistory. It therefore seems likely that the deliberate inhalation of smoke is a very old habit indeed and may even be as old as the human control of fire.

Chapter 10

Pyrotechnology

The use of fire is something so basic and fundamental to human culture that it often seems almost natural. This is partly due to the great antiquity of its use, which has led us almost to think that it was always there and therefore a fact of nature not culture. This is, of course, wrong. For there was a time when our hominid ancestors had not tamed fire and could only make use of it sporadically when naturally occurring fires presented the opportunity. Such serendipitous encounters with fire can have had only a marginal effect on the lifestyle of these hominids; at this stage it was not a central part of their technology. Marx and Engels were undoubtedly right in describing the domestication of fire and the harnessing of its energy for human ends as a 'stupendous discovery, almost immeasurable in its significance'. In this chapter I will show how many of the technological breakthroughs of humankind that are often seen to be the dramatic innovations of a later age have roots that can be traced back to the Old Stone Age. The role of Prometheus, who in Greek mythology is accredited with stealing fire from heaven and teaching the arts and crafts to humanity, is very much like that of Stone Age man, who learned how to control and use fire to develop technologies that built the foundations on which subsequent developments were based. The pyrotechnology, or fire-using technology, of the Stone Age shows that both experimental curiosity and the spirit of invention were as animated then as they are today.

When the domestication of fire began is still a matter of considerable controversy. This is in large part due to disagreements over what can be viewed as adequate proof of the controlled use of fire by hominids. There was a time when the use of fire was thought to have been among the capacities of the earliest kind of hominid, *Australopithecus*, between two and three million years ago. The Australian anatomist Raymond Dart interpreted blackened bones discovered in a cave at Makapansgat in South Africa as proof of this, and consequently named the fossil remains that he found there *Australopithecus prometheus*. Although this was subsequently shown to be based on wishful thinking rather than hard evidence, there are many who still claim very early dates for fire use by hominids. Current opinions range from about 1.7 million years ago down to far more conservative estimates that challenge the notion that even the Neanderthals did not have a straightforward mastery of fire. This will give the reader some idea just how much disagreement there is concerning one of the most important technological breakthroughs of them all. The matter becomes further entangled when we remember that the use of fire, the control of fire and the making and domestication of fire are not simply different ways of expressing the same thing. It seems eminently reasonable to presume – as I have above – that the earliest interactions with fire were of the nature of chance encounters when early hominids may have seized the opportunities that naturally occurring fires presented. It was only subsequently that this became an integral part of life. Very early dates for the use of fire have been proposed for sites in Africa, China and elsewhere.

At least 13 African sites provide early evidence for the presence of fire in archaeological contexts, but this does not automatically prove that fire was used at all by those hominids who used these sites. The action of naturally occurring bush fires or fires that were due to lightning strikes could easily be the direct cause of what could be too readily put down to the deliberate action of hominids. The thermal alteration of tools and bones and the occurrence of burnt sediments need not be the consequence of campfires at all, but can be attributed to natural agencies. This does not mean that all the early cases of apparent fire use in Africa and elsewhere can be extinguished on these grounds; it just shows that we cannot jump to the conclusion that hominid fire use was definitely involved. Some cases are more likely than others, and I shall briefly mention just one such example from Africa. The archaeologist John Gowlett and his associates have put forward the suggestion that fire use is clearly indicated at the site of Chesowanja in Kenya. This site, which has been dated to about 1.42 million years ago, contains animal bones found in conjunction with Oldowan tools and burned clay, all of which they see as the result of the activities of *Homo erectus*. Over 50 pieces of burnt clay were found and the researchers were convinced that in this instance the fire was down to this hominid and not to nature. The arrangement of some stones at the site was interpreted as the remains of a hearth, and many of the larger lumps of clay were found with this hearth. Among the reasons put forward to back up this statement was the fact that the burnt clay was found in direct association with both the artefacts and burnt bones but was not found outside the area of hominid activity. Clearly if the evidence of burning in this archaeological site was put down to the activity of a natural fire then there should have been signs that it had also scorched the area directly beyond the site itself, which it apparently did not. Tests on one of the pieces of clay showed that it had been fired to about 400°C, which belongs to a level of temperatures that are typical for campfires. Both sides in the debate – those for and against accepting the evidence for fire use from this and other sites – have used arguments which are difficult to substantiate. More recently John Gowlett and his Liverpool University team have found what they consider to be the remains of a hearth used for cooking at a site in a forest near Bury St Edmunds in Suffolk, England. Estimated to date from about 400,000 BP, this discovery is likely to generate considerable scepticism among those who believe that fire was not used in such a way in these remote times.

Sometimes those seeking to show that the various possible cases of fire use can be explained away by natural causes have used arguments which can be seen as equally and in some instances even more tenuous than those they oppose. Whilst no case from this very early period can be said to be watertight, the number of sites at which artefacts have been found in conjunction with traces of fire are sufficient in number to have made Glyn Isaac, one of the great authorities on the Lower Palaeolithic archaeology of Africa, express the opinion that control of fire is probably indicated from the Lower Pleistocene onwards.

Probably the most famous archaeological site associated with the origins of the use of fire is that of Zhoukoudian (formerly spelt Choukoutien), not far from Beijing in China, where *Homo erectus* is widely believed to have used fire some half a million years ago. Here the case is seen to be particularly strong, as burnt bones and stones, ash and charcoal have been found in each and every layer of the site. Lewis Binford and Chuan Kun Ho have attempted to persuade their colleagues that this classic case of fire use can be explained more than adequately by natural factors. Binford is known for 'rubbishing' all kinds of claims concerning the capabilities of hominids, and this is just one of many examples of his fundamental approach. His ideas have had a huge impact on archaeology and have caused many re-evaluations to be made. Nevertheless, although some archaeologists have accepted his stamping on the Zhoukoudian fires – in some cases, one suspects, purely on the strength of his reputation – specialists on the Palaeolithic in the Far East have been highly critical. Among them is Geoffrey Pope, who is dismissive of:

> The Binfordian hypothesis of minimal cultural capacities for Pleistocene hominids – imposing impossibly rigorous standards of evidence on archaeological assemblages and postulating elaborate natural alternatives (lightning-caused cave fires, spontaneous combustion, chemical staining, etc.) to explain phenomena most parsimoniously understood as the result of hominid activity.

Binford and his followers have certainly not deterred Chinese archaeologists from making claims for a much earlier cultural development involving the use of fire. In fact the earliest archaeological site in China has been reported as showing the hominid capacity for fire use. The Xihoudu site in Shanxi Province is estimated to be between 1.8 and 1 million years old. Stone tools (and possibly antler tools, although these are more controversial) have been found at the site in conjunction with burnt animal remains including deer bones and horse teeth. In commenting on these finds, Jia Lanpo thinks it is unlikely that the burning was due to a wild fire and so sees it as indicating hominid activity. Other later sites in China also contain charred bones and have been brought forward as further indications that the use of fire is indeed of great antiquity.

The use of fire by Neanderthals is less controversial. Recent reports of excavations undertaken at the Grotte XVI cave site in the Dordogne, France, provide conclusive evidence that Neanderthals who used the cave made fires. The dating of burnt sediments taken from these fireplaces shows that they were in use around 60,000 years ago. The Neanderthals seem to have deliberately chosen lichen as their fuel. They must have collected and dried it in sufficient quantities to make it a practical source. Experiments with burning lichen by the investigators of the site have shown that it is an adequate fuel, giving off fairly high heat. At this site there was no evidence that any cooking took place and it is assumed that fires were made primarily to provide heat and light. The indications are that the cave was used only as an occasional haven from the outside world, as signs of long-term use were lacking. Compared to the rock-lined hearths and excavated fire pits that were made by Upper Palaeolithic people, these fireplaces are rather basic.

Another discovery that may have major implications for our understanding of the Neanderthal mastery of fire is in many ways even more exciting. In 1990 a French caver discovered a cave at Bruniquel in southern France. Soon after this a team of French archaeologists led by François Rouzaud began investigating this deep cavern. Several hundred metres from the entrance a rectangular structure (about 4 × 5 metres in size) constructed out of stalactites and stalagmites awaited them. Based on the dating of burnt bones found within the confines of the structure, its age has been estimated to be at least 47,600 BP. Before this unique structure was discovered, deep caves were thought not to be occupied until the Upper Palaeolithic period, the reason being that since there is no natural source of light at such depths it was believed that without some fairly sophisticated control of fire there would be no way of earlier humans exploring in pitch-black conditions. Both the structure found deep inside the cave and the burnt bones strongly suggest that fire, perhaps even lamps or torches, must have been used as a source of light by whoever visited the caves. The dating places it firmly in the time of the Neanderthals. Investigations of this site are still continuing and it is hoped that much more information will be forthcoming which will shed further light on the capacities and activities of the Neanderthals who appear to have frequented it.

In Europe, and especially in France, objects reported to be lamps have been assigned to all periods of the Upper Palaeolithic from the Aurignacian onwards. The artefacts from the Aurignacian period are considered to be doubtful, and according to Sophie de Beaune, who has made a special study of the lamps, the earliest objects which can be indisputably identified as lamps date to the late Perigordian period. The known number and variety of lamps increases quite dramatically in the Magdalenian, which probably reflects both increased usage during this time and also the fact that later examples are more likely to survive. The vast majority of lamps have been found in France, with rare finds in Spain, Germany and Czechoslavakia. Somewhat surprisingly, the large majority have been found not in deep caves where one might expect them to be more common but in open-air sites, shallow parts of caves and rock shelters. Most lamps are of fairly basic manufacture but a few have carved handles. It is not the case that there is a simple evolutionary sequence from crude lamps to more finely made examples, as the simple and more complex types co-existed, with the former being more common.

Beaune has suggested that the handled lamps might have had a ceremonial use, but this is by no means clear. Although she rejects interpretations of these particular lamps as incense or perfume burners, she notes that laboratory analysis of the organic traces on some of the lamps revealed that wicks were made of a number of substances, including juniper, conifers, an unspecified type of grass and other non-woody residues. When it is burned, juniper gives off a fragrance which has made it a popular incense in historical times in many parts of the world. This may indicate that its fragrant qualities did not go unnoticed in Upper Palaeolithic times. Beaune also conducted her own experiments with wicks of various materials in order to find out which of them was the most efficient and thus more likely to have been used in the Ice Age lamps. Lichen and moss came out on top as the easiest to use. Lichen was also the most popular wick material of the Eskimos in the making of their lamps. The possibility that lichen may have been used in the prehistoric lamps provides an interesting link with the Neanderthal use of it as a fuel.

Clearly the domestication of fire had benefits beyond those of providing heat and light. Without it cooking could not, of course, take place. This did not just herald the primeval beginnings of the culinary arts but also vastly increased the range and variety of plant foods that were inedible unless cooked, many legumes and cereals among them. We can only speculate as to whether or not the use of fire – and particularly the kindling and maintenance of fire – was accompanied by ritual activities and made the subject of mythology by Stone Age man. We can, however, be certain that fire was certainly revered as far back as records go, and that the Greek myth concerning Prometheus was only one of many such sacred stories on the mysterious origins of fire. The ancient Indo-Iranian peoples venerated fire from prehistoric times, as is made clear by the references to it in the earliest holy books of the Indians and the Iranians. Followers of the Zoroastrian religion of Iran have traditionally kept an ever-burning hearth fire, a practice which clearly has its roots way back in prehistory when their nomadic forebears transported embers across the steppes of their ancient homeland.

The historian of religion Mircea Eliade has shown how throughout the world shamans, potters, smiths and alchemists have all been portrayed as 'masters of fire'. The power of fire to transform clay in the case of the potter and metals in the case of the smith was seen not just as a technological accomplishment but as an activity imbued with magic. In the next chapter I shall show how the industrial arts of both the potter and the smith developed out of an already sophisticated technology that involved the use of fire. Perhaps this primary industry of fire was also surrounded by the cloak of magic. Whilst early pyrotechnical activities may well have had their ritual aspects, this in no way diminishes the technological achievements made in prehistory, as Theodore Wertime makes clear:

> Stone Age men were using fire in manifold versatile ways from 25,000 years ago. These ways ranged from the tempering of wood spear points, the oxidation of such pigments as ochre, the annealing of stone projectile points, the fire-setting of quarries (from which the lime kiln and possibly the metallurgic furnace emerged), to the development of the technologies of the cooking hearth. When the full-scale baking of earths for ceramic, bonding, vitric, or metallurgic purposes began some 8,000 to 10,000 years ago, artisans were already launched full-scale into the chemistry and physics of materials . . . One can no longer see pyrotechnology as the movement from the simple to the complex, but as the separation of single strands of new arts and crafts (such as glazes), from complex skeins of existent fire-evolved chemistries and physics.

The numerous activities mentioned in this passage by one of the leading lights in the reconstruction of the early stages of technology give us some understanding that the prehistoric period was far from inert in this sphere of human life. In fact we can see how vibrant and lively was the technological spirit in these early times and how patient and observant its practitioners must have been in pursuit of the perfection of their various industries. Some of the developments sketched out by Wertime will be dealt with in later chapters of this book. In the present chapter I shall concentrate on the heat treatment of various kinds of stone used in the manufacture of tools, in order not only to show the capacities of prehistoric technicians but also to illustrate the considerable legacy they left for later developments in pyrotechnology, particularly pottery and metallurgy.

Modern investigations into the archaic art of applying heat to flint and other materials used in the making of stone tools were hampered in the early days by highly misleading travellers' tales that purported to be eye-witness accounts of such activities. Many of the aboriginal cultures of the Americas used such techniques until very recently. One early report of a dubious kind that set a number of researchers thinking was said to be a technique used by the Seri people of Sonora in Mexico to make arrowheads. It consisted basically of heating a flint in a bed of hot coals and then, whilst it was still hot, applying drops of water on to it through reeds of different sizes in order to fashion it into a tool. The final stages, which involved the sharpening, pointing and smoothing off of the arrowhead, were achieved by very fine droplets of water being applied to the surface of the flint from blades of fine grass.

This account seemed extremely far-fetched, particularly as such a time-consuming and delicate process was apparently entirely superfluous; a more conventional way of making an equally acceptable arrowhead – by the use of a hammerstone and a bone flaking tool – would have been far quicker and easier to do. Expert flint knappers who have tried to reproduce such techniques have found them severely lacking. Firstly, putting flint in direct contact with a naked flame usually results in it shattering. Furthermore, applying drops of water usually results in nothing more than the drops evaporating shortly after coming into contact with the surface of the flint. On the odd occasion in which the water does result in removing chips of flint, this cannot be predicted in advance nor can it be controlled in any way. Thus the technique is entirely impractical.

As Jeremiah Epstein has pointed out, although there are a number of early reports that described the real method of heat treatment used in lithic technology, these were largely ignored until the experiments of the most renowned of all modern flint knappers, Don Crabtree, confirmed that they were sound. During the 1930s, whilst working in the Ohio State Museum lithic laboratory, Crabtree was struck by the difference between the appearance of the flint artefacts made by the Indians of the Hopewell culture (which flourished in Ohio from about 200 BC to AD 400) and the local unworked flint of the kind that they were using as their raw material. The Hopewell flint artefacts differed in that they had what Crabtree described as a 'greasy lustre'. He suspected that heat treatment may have been the factor behind this distinctive lustre. From his considerable experience in making tools from a variety of silica materials, he was more than aware that some were easier to work with than others. Obsidian, agate and jasper were comparatively easier to work using the technique known as pressure flaking, whilst chert, flint and other materials presented more of a problem. Heat treatment of these latter silica materials (by applying the heat in an indirect fashion, for example by heating them in a sand bath rather than exposing them directly to a flame) has been shown to make it far more straightforward to work with them. The heating of lithic materials has to be done gradually, as too rapid a change in temperature causes thermal shock, resulting in the shattering of the silica materials; the same also occurs if they are not allowed to cool down slowly after they have been subjected to heat. Crabtree conducted numerous experiments over the years in order to gain a deep understanding of the subtleties of lithic heat treatment, finally publishing these findings in 1964. By trial and error he had been able to work out that:

Silica materials varied considerably in the length of time and amount of heat necessary to bring about the desired change. For each type of silica material there appeared to be a critical temperature range below which, regardless of the time involved, no change would take place, and above which it would crack and craze. On the other hand, some of the minerals

I The Palette of Narmer, Hierakonpolis, Egypt.

II and III The megalithic temple of Hagar Qim, Malta, 3500–3000 BC.

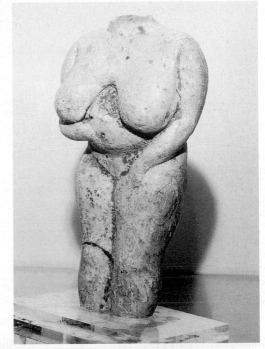

IV The 'Maltese Venus', from
Hagar Qim. Height 12.5 cm.

V The 'Sleeping Lady' from
Hal Saflieni Hypogeum, Malta.
Length 12 cm.

VI Main hall of an underground temple, Hal Saflieni Hypogeum, Malta, 3rd millennium BC.

VII Sculptures at the site of Lepenski Vir, Serbia, c. 6,000 BC.

VIII The 'New Rosetta Stone', Nuzi, Iraq, 2nd millennium BC.

IX 10,000-year-old engraved bones, Remouchamps cave, Belgium.

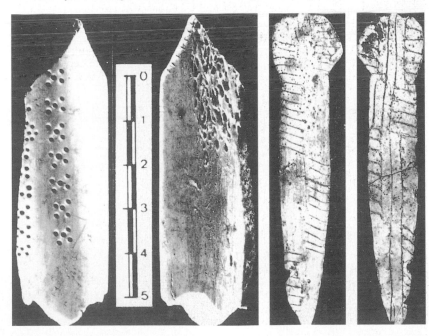

X Scene depicting Neolithic skull surgery (diorama, Science Museum, London).

XI Set of three drills used by Dr T. W. Parry for experiments on Neolithic skull surgery.

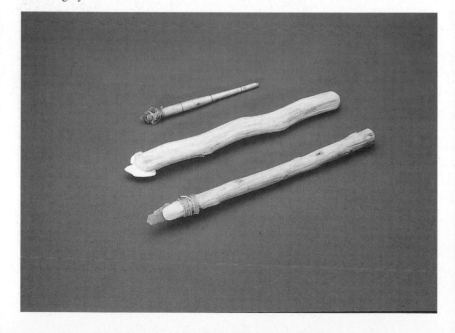

XII Trepanned Gusii man, Kenya.

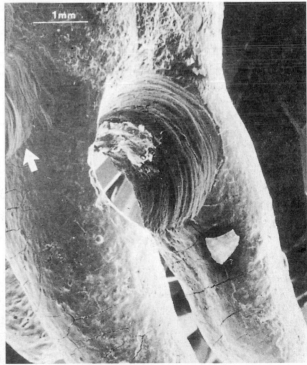

XIII Hole drilled in tooth by Neolithic dentist, Hulbjerg, Denmark.

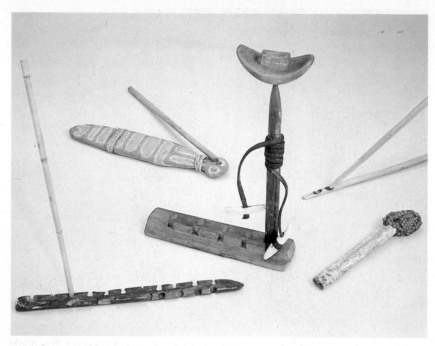

XIV A selection of traditional fire drills.

XV Prehistoric bow drill used for making fire, Kahun, Egypt, 4th millennium BC.

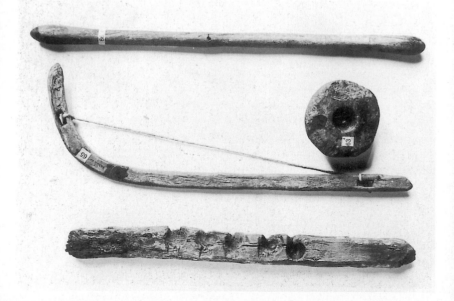

XVI Goddess
figurine, Grimes
Graves, Norfolk,
England. Height
10 cm.

XVII Base of a shaft
in a Neolithic flint
mine, Grimes Graves,
Norfolk, England.

XVIII The 'Venus of Laussel', Dordogne, France (Upper Palaeolithic).
Height 44 cm.

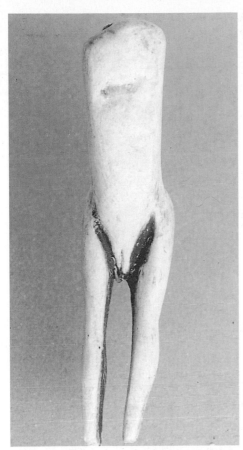

XIX Ivory figurine of young girl, Laugerie Basse, Dordogne, France (Upper Palaeolithic). Height 7.7 cm.

XX Female figurine carved from mammoth tusk held by its discoverer Lev Tarasov, from Gagarino, Russia (Gravettian period, Upper Palaeolithic). Height 12.7 cm.

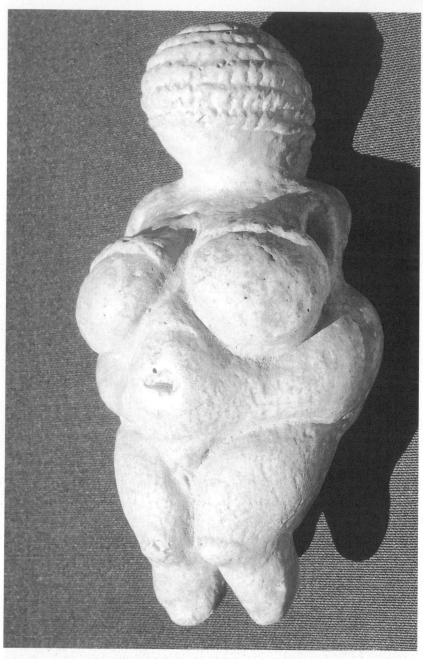

XXI The 'Venus of Willendorf', Willendorf, Austria (Gravettian period, Upper Palaeolithic). Height 11 cm.

XXII Adult male burial, Sungir, Russia (Upper Palaeolithic).

XXIII Spear thrower carved with two fighting bison, Le Mas d'Azil, Ariège, France (Upper Palaeolithic).

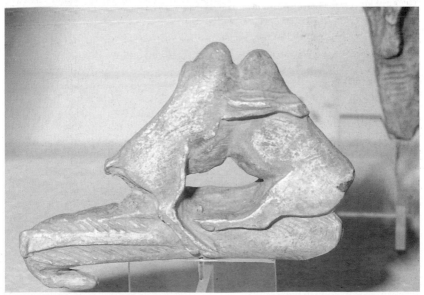

XXIV 30,000-year-old painting of animals, Chauvet cave, Ardèche, France (Aurignacian period, Upper Palaeolithic).

XXV Reconstruction of Neanderthal
scene.

XXVI Hand-axe with fossil shell,
West Totts, Norfolk, England
(Lower Palaeolithic). Length 13.5
cm, Width 7.8 cm.

XXVII 'The Pearl of Yakut Archaeology' – the archaeological site of Diring-Ur'akh, Siberia.

XXVIII 3 million-year-old pebble, Makapansgat cave, northern Transvaal, South Africa. 6cm across.

had to be held in the critical temperature longer than others in order to bring about the desired change.

A leading French specialist on Palaeolithic stone tools, François Bordes, who was also an expert flint knapper, recognised the significance of Crabtree's insight into what was then a little-understood aspect of prehistoric technology. In 1969 Bordes reported that some Upper Palaeolithic artefacts of the Solutrean period (about 21,000 to 18,000 BP) had clearly been subjected to heat treatment. This placed the technological innovations of lithic heat treatment firmly in the Old Stone Age. Crabtree's statement cited above shows that the use of heat in the preparation of raw materials for industrial production was not a straightforward or simple procedure. For Palaeolithic man to have used such pyrotechnology successfully he would have had to master a number of skills. Detailed knowledge of a range of materials, as well as a very accurate sense of timing and temperature control and maintenance, would have been essential. In these three fundamental aspects of Stone Age pyrotechnology one can see key elements that are found in the subsequent industrial activities of firing pottery and smelting metals.

Most of the detailed studies and experiments on lithic heat treatment since the pioneering work of Don Crabtree have been undertaken in North America. As the practice is known to have been widespread in aboriginal cultures, this has also allowed detailed comparisons to be made with the large collections of Indian stone tools and weapons. Barbara Purdy has done a series of experiments on the particular kind of silica mineral known as chert (of which flint is the most well-known variety). She applied heat treatment to chert from Florida and in the light of her findings was able to recognise that similar procedures had been used in the making of numerous prehistoric chert artefacts from the region. Her experiments showed that between 240° C and 260° C the local chert went through a colour change, but was not really altered beyond that until a heat of 350° C had been attained. Having established that this latter temperature was the critical temperature for altering the chert, she then tested the relative strength of heated and unheated chert. Two kinds of strength were examined – compressive strength and point tensile strength. She found that the compressive strength of the heated chert samples was 40 per cent greater than that of the unheated controls, whilst the point tensile strength of the heated chert was 45 per cent *less* than the unheated controls. Although this appears to be a rather paradoxical discovery, it is not, as Purdy explains:

This seeming contradiction is easily explained. The binding of the microcrystals that occurs when the rock is heated adds compressive strength through cohesion. The increase in uniformity that increases strength under compression is the very factor which decreases point tensile strength. The individual microcrystals are bound more firmly together. Therefore, when the flaw is introduced, which is preliminary to and necessary for fracture to occur, failure takes place more readily because the specimen responds more like glass rather than like a rock aggregate.

In other words, altering chert through heat treatment makes it easier to flake and therefore to produce tools of better quality with less effort. The firing of chert alters its structure, making it more like easier-to-work materials such as obsidian (naturally occurring volcanic glass). Heated chert also has the added advantage of providing a sharper cutting edge. The heat treatment seems to have been especially useful in the making of artefacts requiring balance and symmetry, particularly arrowheads and other projectile points. Having established both the means of properly applying heat treatment and the advantages of doing so, Purdy then studied a collection of more than two thousand projectile points found at the prehistoric Senator Edwards Chipped Stone Workshop Site in Marion County, Florida. These artefacts belong to a cultural phase known as the Preceramic Archaic and date between 9000 and 5000 BP. The collection included both finished projectile points and many unfinished ones in various stages of the manufacturing process. She was able to establish by the appearance and colour of the artefacts that at least 750 had been subjected to thermal alteration. Most of the specimens that showed clear signs of heat treatment had already reached a fairly advanced stage of manufacture. This led her to the conclusion that the application of heat did not occur before the manufacturing process began but at some stage during the process. It also became apparent that different kinds of projectile points were given heat treatment at different stages in the manufacturing process.

Clearly there was a sophisticated knowledge on the part of the prehistoric Indians who made these artefacts which gave them the means to take full advantage of the raw materials available to them. The use of heat treatment in lithic technology continued among native Americans into the twentieth century. Reports by anthropologists and travellers include accounts of how aboriginal peoples of the Plains and Plateau regions of America would bury obsidian and other minerals in damp earth for twenty-four hours or more in order to prepare them for manufacturing into tools. Such descriptions have helped archaeologists understand a number of otherwise puzzling finds at prehistoric Indian sites. One example will illustrate the point:

In 1894, Dr J. F. Snyder excavated Mount No. 1 of the Baehr Group, on the west side of the Illinois River, thirteen miles below Beardstown, and opposite the mouth of Indian Creek (Illinois). At the base of the mound was an oval of clay on which was a mass of black hornstone implements, that apparently had been thrown down in lots of 6 to 10, with sand over and between each lot, as though to isolate them from each other. This deposit of 6,199 flints was covered with a stratum of clay, 10 inches in thickness; and on this a fire had been maintained for some time ... the flints forming the nucleus of this mound are very ... rudely fashioned; some are quite neatly finished, but the greater part of them are only chipped and ill-shaped.

Before Crabtree and others had rediscovered the heat treatment aspect of prehistoric lithic technology, archaeologists were somewhat at a loss in trying to explain what such finds meant. Among the suggestions made were that such finds might have been grave goods, a means of hiding artefacts from enemy tribes or simply examples of tools in the making that had been abandoned by their makers for some unknown reason. The last of these suggestions was closest to the mark but the role of heat treatment was, as I have said, not understood at this time. Clearly the site in Illinois indicates the large-scale operation of heat treatment. It has now become clear that the use of heat treatment in aboriginal North America can be traced back to the Palaeo-Indians, who are considered to be the earliest inhabitants of the New World. This, in conjunction with the fact that the practice is known from the Upper Palaeolithic period in France, and has been reported from Aboriginal Australia, South America and Japan, shows that it is both a widely used technique and one of considerable antiquity. Whether heat treatment in lithic technology precedes that of ceramic technology is, at the moment, unclear.

We have seen in an earlier chapter how pottery first arose in Japan and parts of the East Asian mainland about 12,500 years ago, but it is a little-known fact outside of archaeological circles that the invention of pottery is not the earliest evidence for the existence of ceramic technology. The discovery of very early pottery in Japan showed that such a technological accomplishment not only existed long before the advent of farming (pottery was considered to be one of the most important innovations of the Neolithic period) but also attained superb standards in a Mesolithic kind of society. Yet even the precocious nature of East Asian pottery was preceded by an earlier tradition of ceramic technology. There have been a number of reports of the deliberate fire-hardening of clay artefacts from various parts of the world, including not only Japan but also Siberia, Russia, North Africa, the Pyrenees, and earliest of all at Dolni Vestonice and neighbouring sites in Moravia, Czech Republic, that place the known origins of ceramic technology a further 15,000 years or more before the earliest Japanese pottery, way back into the early Upper Palaeolithic period.

The Moravian sites of Dolni Vestonice I and II, Predmosti, Pavlov I and II and Petrkovice have all yielded ceramic artefacts dating from the Gravettian period of the Upper Palaeolithic. The 26,000-year-old ceramic artefacts from Dolni Vestonice are comprised of more than 6,750 fired clay fragments, over half of them pieces of broken animal (and a few human) figurines. About ten years ago this important site was revisited by a team of researchers seeking to understand the technological procedures that had taken place in Upper Palaeolithic times. As Crabtree and Purdy had made experiments with local raw materials in order to understand the role of heat in lithic technology, so the team at Dolni Vestonice (consisting of Jiri Svoboda, Olga Soffer, Bohuslav Klima and Pamela B. Vandiver) sought to understand the ceramic technology of the site by making their own replicas of figurine fragments using the local soil. They also made a replica kiln based on the two structures identified as prehistoric kilns at Dolni Vestonice. Their findings were quite surprising, for although many of the prehistoric ceramic fragments had clearly exploded with a bang as a result of being fired (at temperatures between 500° C and 800° C), this does not appear to be due to a crude state of technological capacities. The investigators found through their own experiments with firing local materials that:

> such properties of the Dolni Vestonice loess [soil] as the near-zero drying and firing shrinkage, low thermal expansion, and relatively high porosity make thermal shock an improbable event during firing. Successful replication of thermal shock required firing of pieces larger than 1 cm that were so wet they would barely hold their shape and rapid placement in the hottest part of the fire. Thermal shock did not occur accidentally but required intentional effort and practice ... We measured success in replicating thermal shock by the production of high-energy, branching fractures; however, the other measure of success, and the one most immediately sensed as the figurines were put in the fire, was the evolution of steam, often with a sizzle. Then the figurine would shatter with a pop, sometimes sending pieces flying through the air. In view of the physical properties of the loess that constrain its behavior and these experimental observations, we believe that intentional effort was required to produce the thermally shocked microstructures found in Dolni Vestonice ceramics.

So, if the people who were firing these figurines had to go out of their way to make sure that they exploded, what were their reasons for doing this? It seems likely that their steering of the available technology was subordinate to another purpose other than the purely practical. It appears that if they had wanted to make and fire figurines for permanent use they could easily have done so. The most likely answer is that the making and almost immediate destruction of the figurines had some ritualistic meaning. Most of the fragments of the figurines were found not in the main settlement area of Dolni Vestonice but at two kilns situated about 80 metres away uphill. The research team thought that this might indicate that the ritual destruction of the figurines may have been undertaken by only a minority of the prehistoric community that dwelt at Dolni Vestonice. If there was such a minority, then they would most likely have been some sort of spiritual practitioners, akin to shamans, priests or priestesses, perhaps 'sacrificing' the figurines in magical hunting rites. What seems certain is that there was no apparent desire to produce permanent 'works of art' *en masse* at Dolni Vestonice and the other Moravian sites.

The ritual use of this ceramic technology has been likened by its investigators to the early prehistoric use of metals; for example, the manipulation of copper (which can be traced back at least 10,000 years in the Near East) was initially undertaken to make ornamental objects and only some 4,000 years later was its use extended to the making of utilitarian artefacts. Thus in the case of both fired clay and copper technology non-utilitarian uses precede its practical applications by millennia. There are both similarities and differences between the lithic and ceramic technologies that have been considered in this chapter. As far as we know, the heat treatment of flint, chert and other such minerals was undertaken for what are essentially utilitarian reasons, for there is no evidence that the deliberate destruction of such materials was ever practised for ritualistic purposes, although it is not inconceivable that it may have been done at some time in the past. Whilst the ceramic technology at Dolni Vestonice was non-utilitarian in its character, it did, like lithic heat treatment, require a body of practical knowledge to achieve its very different goals. In the light of what has been said above, the development of pottery should not be seen as an innovation that was without precedents. It should rather be seen as the application of ceramic technology to the making of vessels and containers (which had previously been made from a variety of other materials such as stone, leather, bark and, of course, in the form of baskets from which many pottery forms clearly derive not only their shape but also their decoration), and as such it is the merging of two technological traditions that preceded it.

Chapter 11

Back to the Grindstone

The use of containers of some form or another is undoubtedly a very early form of material culture, although the study of such objects in the Palaeolithic period has been rather neglected. One of the reasons for this lies in the fact that baskets, leather buckets and other vessels that were made from perishable organic materials simply do not survive for thousands of years in anything like the number that ceramic artefacts do. Although the open-air site of Pavlov I in Moravia (not far from Dolni Vestonice) was excavated in the 1950s, it was only very recently that a re-examination of ceramic fragments from the site by archaeologist Olga Soffer revealed that impressions of interlaced fibres were present on a few of these clay specimens. Four pieces of fired clay have negative impressions on them which show that they were, about 26,000 years ago, in direct contact with some kind of textile or basketry. The study team that investigated these fortuitous discoveries concluded on the basis of their analysis that:

Since the warps and wefts are produced by essentially cordage production techniques, it may ... be confidently assumed that both string and rope were in use at Pavlov I ... the Pavlov I impressions represent technically well made items; they are scarcely 'primary essays in the craft'. Further, the general regularity and relatively narrow gauge of the elements used in [these] specimens suggest considerable antecedent development not only for these specific techniques but also for the perishable industry or industries at large ... if they are portions of bags or mats, they may have served the 'usual' role of these items – that is, as flooring or sitting/sleeping platforms in the case of mats, and as storage/transportation devices in the case of bags. If, on the other hand, they are portions of cloth fabrics, they could represent blankets or items of clothing such as shawls, shirts, skirts, sash fragments, etc. ... the perishable impressions from Pavlov I could represent the intentional application of clay to the 'outside' of some flexible containers to provide a simple form of mould not unlike that evidenced in some early North American ceramic production.

If clay were deliberately applied to baskets or other containers then this would suggest a very early foreshadowing of the earliest pottery tradition in the world, that of the Jomon culture of Japan about 13,000 years ago – for the textile impressions on clay from Pavlov I are twice as old as this, dated to 26,000 BP. These textile impressions are easily the oldest known evidence for the existence of a fibre-based technology in the world but, as the team said, the textiles that were being made at this time were obviously made by people drawing on a well-established technology rather than undertaking the first fumbling attempts to make cloth or baskets. This makes it clear that textile technology must have started considerably earlier. Although the specimens from Pavlov I could, technically speaking, have been produced without the use of a loom, it is quite likely that some kind of frame was used to make them.

It is only in exceptional circumstances that evidence for the early use of such perishable organic materials can be found. So although the evidence for the use of vessels and containers in pottery-using cultures is abundant and the evidence for the same in pre-pottery cultures is paltry, we should not, of course, presume that the latter made little use of such things. Indeed it is obvious that no culture can go without containers in which to carry and store items. Mention has been made in the previous chapter of the lamps of the Upper Palaeolithic period which represent a specific type of container, but they are hardly the only type known from this time. Most containers that have survived from this period are made of stone, such as the oval limestone receptacle from the Aurignacian period found at La Ferrassie in the Dordogne, France, by Denis Peyrony. He suggested that it might have been used as a mortar, but subsequent examination of it has demonstrated that there are no traces of percussion wear that such a robust usage would inevitably have left on the object. There are indications that the rim of the receptacle was artificially shaped, but there is no way of inferring what it contained. There are many uses for such a container that would leave no traces for the archaeologist; whether it stored food, water or something else simply cannot be known.

There are many other objects in the Palaeolithic archaeological record that could be vessels of some kind. It has been suggested that parts of human skulls found at a number of Upper Palaeolithic sites (including Dolni Vestonice) could have acted as drinking cups. Although there are many known instances of such practices from more recent times – for example, the use of skulls as ritual drinking vessels among the Tibetans and their use to transport water by the Aboriginal Tasmanians – there is no way to demonstrate that Palaeolithic peoples ever made use of skulls in this way. The object held in the hand of the 'Venus of Laussel' (see *Plate XVIII*) has been identified as a drinking horn and even as evidence for the knowledge of fermented alcoholic drinks! This object is open to a number of other interpretations, such as Marshack's belief that it relates to observations of the moon and Dirk Huyge's suggestion that it is a musical instrument (see Chapter 15). Animal horn certainly provides a ready-made vessel which *could* have been used in Palaeolithic times (as it has in historical times), but to leap to the conclusion that alcohol was involved is pure conjecture. A limestone rock, also from Laussel, has been identified as a container for ochre, an iron ore greatly prized in the Palaeolithic period (see Chapter 13).

At the Magdalenian site of Hostim in central Bohemia, the archaeologist Slavomil Vencl discovered 48 pieces of sandstone and other stone which he identified as the fragmentary remains of more than 27 bowls. He also found quartzite and other raw material fragments of 11 dishes. Yet the number of such containers found at Upper Palaeolithic sites is rather small, which may partly reflect that this kind of object has gone largely unnoticed by archaeologists due to the minimal — if any — evidence of human workmanship. It is also probably the case that most containers from the Upper Palaeolithic period were made from perishable materials, and many were naturally occurring items such as large leaves, bamboo, bottle gourds, ostrich egg shells, animal stomachs and so on which would also have required little or no human modification to make them useful. Such objects, even if they survived, would rarely show clear signs that they had been used as vessels or containers.

Bearing in mind the poor evidence for the use of receptacles in the Upper Palaeolithic, it is hardly surprising to find little indication of the use of such artefacts from earlier periods. A receptacle containing powdered ochre found at the Grotte de Néron site at Soyons, Ardèche, France, cited by Sophie de Beaune as indicating the existence of Middle Palaeolithic containers, is one of very few reported cases of such artefacts. Nevertheless, the existence of containers may be inferred even from the Lower Palaeolithic period without direct evidence. The amateur archaeologist R.J. MacRae has studied the availability of raw material for the making of flint tools that existed in the Lower Palaeolithic period in parts of Oxfordshire, England. Having found that both the raw material itself, in the form of unworked flint nodules, and ready-made tools must have been transported over considerable distances, he questions whether Lower Palaeolithic man might have had some more convenient way of transporting flint than simply carrying it in his hands. Not only would the manually held tool or nodule be a tedious means of transport, it would also prevent the hands from performing other tasks whilst travelling. It seems probable that Lower Palaeolithic man had the intelligence to fashion some suitable alternative; MacRae suggests that a basket, a rawhide sling, a fibre belt or an animal-skin shoulder-bag could have been used in these times.

A similar case has been made for the use of the baby-sling in remote times, by Tim Taylor, again without the availability of direct evidence. He suggests that whilst a baby chimpanzee clings on to its mother's fur and so leaves her limbs free for other tasks, the early human infant (whose mother was standing upright and had lost her body fur) would have been an intolerable burden without the invention of the baby-sling, which he describes as a female invention and 'the first characteristically human artifact'. Such a sling could have been a simple animal-skin shoulder-bag, similar to that proposed by MacRae for transporting flint. Thus the existence of simple man-made (or woman-made) containers seems to be not just an innovation that would have made life easier for Lower Palaeolithic people but one that was essential to their very existence.

The people of the Old Stone Age made limited use of mechanical contrivances, not out of a lack of inventiveness nor because they lacked the necessary intellectual qualities. Their hunting and gathering lifestyle normally required a high degree of mobility which would have made it highly impractical for them to tie themselves down to the making and maintenance of machinery. As a general rule – although there were some exceptions to this – if an object were not portable it was useless to them. Basic mechanical operations were largely limited to the realms of food-processing and food-procurement – that is to say to the crushing, pounding and grinding implements used in preparing mainly plant foods (and also to process pigments), to fire-making, and to the improvement of weapons used in hunting animals. I will deal with each of these three in turn.

A great deal of emphasis has been placed on the role of hunting in the Old Stone Age, and the gathering of plant foods has far less attention than it deserves. The detailed study of modern hunter-gatherers by anthropologists has shown that gathered foods (particularly edible plants but also eggs and sessile life-forms such as shellfish) often make up 80 per cent of a community's diet, with meat derived from hunting making up the rest. Hunting has been given an inordinate status both by archaeologists and by hunting peoples themselves. In the latter case this is largely due to the fact that it is the men who do most of the hunting and they therefore pride themselves on their own achievements in providing meat and tend to play down the considerable contribution of women. Prehistorians are presented with an archaeological record that contains far more information on hunting than on gathering activities. Animal bones are often found in abundance at Stone Age sites, whereas plant remains and other evidence of the gathering aspect of the economy are sparse, owing to the poor survival of botanical specimens. This has understandably led to a concentration on the available data, which can give the false impression that hunting was the most important means of getting food in the Palaeolithic period.

However, there are a number of kinds of Palaeolithic artefacts that show that food-processing was an important aspect of life for people long before the first settled farming communities of the Neolithic period. Interest in such pre-agricultural tools by archaeologists has on the whole been limited to their presence in the Natufian culture of the Levant (*c.* 10,500 to 8000 BC), because it is in this area that agriculture made a very early appearance on the world stage. The Natufian culture, although belonging to the end of the Upper Palaeolithic period, had already attained many of the features that were typical of subsequent farming cultures of the Neolithic – the use of the mortar and pestle for processing cereals, sickles for harvesting, living in permanent settlements and storing their food surpluses. Obviously the study of the Natufian culture is intimately involved with the emergence of agriculture in this part of the world, but implements used in the preparation and processing of foods have been found far away from the Levant and can be traced way back into the Palaeolithic period; thus they cannot be seen simply as setting the scene for agriculture but must have been an integral part of the hunter-gatherer lifestyle. The electric food-processors of today are the latest outgrowth of a technological tradition that goes back to the Old Stone Age methods of crushing, pounding and grinding foodstuffs.

Even today the grindstone is still used as a symbol of hard and repetitive work. For us the phrases 'back to the grindstone' and 'the daily grind' are merely metaphors (although in many instances people's daily work is as repetitive and arduous as it was in prehistoric times, despite the assumption that modern technology has liberated us), but for countless generations, the grinding of corn and other cereals was a daily task. The processes of pounding and grinding both require the use of an upper and a lower stone. A number of different names have been given to both kinds of stone used in both operations. In the process of pounding, the lower stone is known as an anvil or a mortar whilst the upper stone is called a pounder, a pestle, a hammer-anvil or a percussion muller. In grinding, the lower stone is called a quern, a milling stone, a metate (a term usually limited to the New World), a saddle-quern or simply a grinding stone; the upper stone is a muller, a rubber, a mealing stone or a grinder, or in the New World it is referred to as a mano.

The evidence for the use of such simple food-processing tools can be traced far back into prehistory. American archaeologist George Carter has claimed that grinding artefacts in the form of both manos and metates that are at least about 80,000 years old have been found in California, although this is widely rejected simply because most archaeologists do not consider that there is sufficient evidence for the populating of the New World until 12,000 BP or slightly earlier. Stone artefacts that show evidence of use wear caused by the grinding process were found at Florisbad, Orange Free State, South Africa, and have been dated to 48,900 BP. Grinding stones from the Bushman Rock Shelter in eastern Transvaal have been dated to approximately 43,000 to 47,000 years old and were used to process ochre. At the Olieboomspoort Cave in the Transvaal, five grindstones, also utilised in ochre preparation, are about 33,000 years old. Recent finds of 30,000-year-old grinding stones from Cuddie Springs, New South Wales, Australia, by Richard Fullagar and Judith Furby demonstrate the early use of such plant-processing equipment on the continent; these kinds of artefacts have been used by modern Aboriginal groups, as will be described below. A microscopic analysis of the Cuddie Springs grinding stones indicates that they were used to process seeds. Grinders and pestles from the Mousterian levels at the Molodova I site in the Ukraine are more than 44,000 years old, and similar artefacts from the nearby site of Molodova V are at least 40,000 years old. Large grindstones most probably used in food-processing and of

Mousterian age have also been discovered at the Cueva del Castillo cave site in Spain.

These finds from Africa, Europe and Australia indicate that such pre-agricultural tools are of considerable antiquity in diverse parts of the world way before the Neolithic period. It is likely that numerous upper and lower grinding and pounding stones have been disregarded or at best unreported by archaeologists more interested in the collecting and describing of tools associated with the more glamorous and male-oriented occupation of hunting. This has, in turn, contributed to the overall impression that food-processing tools were a minor component in early prehistoric technology. The importance of male tools used in hunting and butchery is over-emphasised at the expense of the tools that were most likely to have been used by women. Whilst there is evidence for the grinding process going back about 50,000 years, it seems that pounding is a much earlier process. Nancy Kraybill has suggested that various stone artefacts from Lower Palaeolithic sites – among them Olduvai Gorge – are to be identified as pounding stones, and perhaps even in these primeval times some sexual division of labour existed in which there was a clear distinction between male and female technologies, the one based on hunting and the other based on gathering. Before considering aspects of the hunting weapons used in the Old Stone Age, I will first describe the ways in which early humans engendered that most vital of resources, namely fire.

There are basically two ways to make fire that were available before the invention of the match in the nineteenth century – by percussion or by wood-friction. The first method involves the making of sparks by striking a flint or other stone (such as chert or quartz) against iron pyrites. In later times iron, and then steel, replaced the iron pyrites. In order for this method to be effective, it is crucial that the sparks fall upon a kind of material that is easily kindled. Among the most favoured of such tinder materials are: punk, also known as touchwood, which is wood that has become rotten due to the presence of fungus; one of a number of puffballs; and the polypore fungus *Fomes fomentarius*. The last of these has been found at the eighth millennium BC Mesolithic site of Star Carr in Yorkshire, England, and its presence is probably due to its use in fire-making. More direct evidence of the use of the percussion method of making fire also comes from Yorkshire in the form of a flint found in conjunction with a nodule of iron pyrites from the Neolithic site at Rudstone.

Although there are a number of distinct methods of making fire by wood-friction, the principle that underlies them all is essentially the same. The friction caused when two pieces of wood are rubbed together results in the production of a little pile of wood-dust which smoulders and can be blown upon to make it glow, at which point it can ignite the tinder. As in the case of grinding and pounding procedures, the production of fire by wood-friction involves an upper and lower piece. The lower piece of wood is stationary and called the hearth, whilst the upper one is called the saw, plough or drill depending on which process is involved. The fire-saw works as follows. Typically a slit is cut across the underside of the hearth and the saw is rapidly moved across the slit repeatedly until the hot dust sets fire to the tinder placed directly below the hearth. In certain Aboriginal cultures of Australia where a highly portable technology was essential to their mobile lifestyle, a single artefact often played numerous roles depending on its context. A makeshift fire-saw was made by using a shield as the hearth and a spear thrower (for which see pp. 164–5) as the saw. The fire-plough works by rubbing *with* the grain of the hearth rather than against it as is the case with the fire-saw; in other respects the procedures are very similar. The fire-drill has a number of forms, the most simple of which involves the holding and rotating of a vertical drill stick in the hands whilst simultaneously applying downwards pressure on it to create friction with an indentation (or pit) in the hearth. A notch is often made at the side of the pit in order for the wood-dust to be released on to the tinder material, and many such fire-drills have hearths with multiple pits and notches.

All the methods described so far involve simple muscle power to engender sufficient friction to make a fire, whereas the other variants of the drill – the thong-drill, the bow-drill and the pump-drill – all provide mechanical assistance to the user (see Plates XIV and XV). As the third of these was rarely used in fire-making, I shall not describe its workings here. The way the first of these operates is described by H.S. Harrison:

> The thong-drill is rotated by a cord passed round it in a simple loop, the two ends usually having wooden or bone handles for convenience of working. The two hands of the operator pull on the thong in such a way that the stick repeatedly changes its direction of rotation as the hands move to and fro. Obviously there is here a necessity for the drill to be held upright in firm contact with the hearth by pressure from above, and a small socketed holder of wood, bone, or stone, or even the cut end of a coconut shell, is provided for this purpose. This socket-piece may be held down on the top of the drill by an assistant, or if its shape is suitable, as it usually is in the Eskimo appliance, it may be gripped in the mouth of the fire-maker. The thong-drill is used especially by the Eskimo, but is found also in northern Asia, and here and there in India and Indonesia.

The bow-drill works in a similar fashion to the thong-drill except that the two ends of the thong are attached to one another. The thong is fastened at both ends to a bow and is loose enough to allow it to be looped around the drill. The advantage of the bow-drill over the thong-drill is that only one hand is required to move the thong, thus leaving the other hand free to grip the socket-piece or hand-rest. This device has been used by cultures across the world from the Eskimo of the arctic to the Aborigines of Australia, and was even used among the Ancient Egyptians, as is known from a perfectly intact bow-drill found in the tomb of Tutankhamun. It is not only this wide

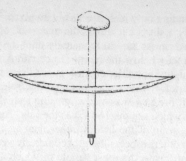

Figure 26

distribution of the bow-drill that suggests that it was an invention of the Stone Age period, for a few of these artefacts have survived from this time. *Figure 26* shows a bow-drill with a hand-rest made of stone, a type which was known in the Maglemosian culture of Scandinavia during the Mesolithic period. The drilling artefacts found in Maglemosian sites include a stone hand-rest, an antler hand-rest and a bow made from a rib. But the abundant finds of nodules of iron pyrites in such sites make it probable that the most common means of fire-making was by percussion and that the bow-drill was generally used to drill holes for various practical and ornamental purposes rather than to make fire. Although there are comparatively few surviving drills from Stone Age sites, the large number of objects that have been found with drill-holes in them shows us that the practice of drilling was commonplace even at this time.

Unfortunately, due to the ravages of time, which have obliterated most of the material culture of earlier epochs, many aspects of the technology of early man are known only from the few fortuitously preserved artefacts, and we may presume that other developments of these remote times have left no tangible traces at all. Stone tools have survived in their millions but we know, for instance, that wood was also used from the Lower Palaeolithic period onwards, although the surviving examples are few and far between for the simple reason that organic materials such as wood do not have the almost indestructible nature of stone. The use of wood has been demonstrated indirectly by the microscopic analysis of stone tools dating from this time. Lawrence Keeley was able to show that when stone tools are looked at under the microscope they show distinctive kinds of polish which are invisible in almost all cases to the naked eye. By comparing the various kinds of polish produced when modern reproductions of ancient stone tools were used on a variety of materials such as leather, plant fibres and of course wood, it became possible to match up the distinctive kinds of polish on these contemporary tools with the originals. By this kind of study, known as microwear analysis, it is possible, at least in some cases, to say what kind of materials were worked on by a particular stone tool, even if it is of Lower Palaeolithic age.

More direct evidence in the form of actual specimens of wooden artefacts has been found in a number of parts of the world. What is certainly one of the oldest wooden artefacts discovered so far is the Clacton spear point found in 1911 in Clacton, Essex, England, and dated to between 420,000 and 360,000 years old. When the ability of early man to be a successful hunter was brought into question by Binford and others, who suggested that the human career consisted of an extremely long apprenticeship as a scavenger (having the ignoble status of competing with hyenas and the like for their daily meat), the Clacton spear point was reinterpreted as either a digging stick or a snow probe, it being thought that hunting spears did not exist at such an early date. Now, however, it seems safe to return to the original identification of it as a spear due to an analogous discovery by Hartmut Thieme in a coal mine at Schöningen in Germany of three javelin-like spears. These spears are thought to date back as far as 400,000 years ago, and it appears that they are throwing, rather than thrusting, spears; the longest of the three is 2.3 metres. Recent analysis of a circular hole in a half-a-million-year-old horse scapula from the Boxgrove site in Sussex, England, by forensic pathologist Sir Bernard Knight indicates that it may have been made by a spear, perhaps a throwing spear. If this hole was made by a spear, then this would be evidence of the use of such hunting weapons about 100,000 years before both the Clacton and the Schöningen finds.

Hunting weapons are not the only wooden artefacts to survive from these remote times. At the Palaeolithic site of Nishiyagi near Akashi in Japan, a mulberry wood artefact that is somewhere between 50,000 and 70,000 years old has been identified as a plank, although its exact purpose remains obscure. Even more remarkable is an extremely early find from the Acheulian site of Gesher Benot Ya'aqov in the Jordan Valley in Israel. A willow plank that has clear signs of having been deliberately polished on one of its surfaces was discovered at this site in 1989. Naama Goren-Inbar, Ella Werker and S. Belitzky, the team that have analysed this artefact, estimate its age to be somewhere between 750,000 and 240,000 BP. Even if it belonged to the later end of this range (which would make it younger than the Clacton and Schöningen finds), its significance would remain largely undiminished. The polishing of wood may seem a trivial matter in the history of technology but when we consider the extremely early period to which this artefact belongs, its discovery takes on a far greater significance. It provides a tantalising glimpse of a technological tradition of which we know so little, and one can only agree with the comment of those who found it: 'it is possible that we have underestimated the capacities of the Middle Pleistocene hominids and further "unconventional" discoveries may cause us to revise our opinion of their abilities'. Whilst the use of wood and other perishable materials no doubt played a large role in the practical activities of early man, the inferences that can be drawn from the few surviving traces of these largely lost technological traditions are limited.

The invention of the spear thrower in the later part of the Upper Palaeolithic period (perhaps much earlier) meant that both the skill and physical strength required to throw a spear were considerably reduced. This mechanical aid is a stick, usually under a metre in length, with a peg or hook at one end which fits into a hole in the butt end of the spear and a finger hole at the other end. The spear thrower acts like an extension to the throwing arm and so increases the leverage available and reduces the amount of wrist action required. Although the spear thrower is, of course, primarily a practical device, some of the Upper Palaeolithic examples are highly decorated. Some of the artistic designs are completely unique, and in the case of a particular Magdalenian example from Le Mas d'Azil, Ariège, France, quite bizarre. This spear thrower (see illustration on back cover) is made from antler and has been finely worked. It shows a young ibex in the process of defecating, with two birds perched upon the faeces! Another decorated example was also found at this site (see Plate XXIII). The use of this device continued into modern times among tribal peoples in many parts of the world, including South America, New Guinea and Australia. Brian Cotterell and Johan Kamminga, who have researched the mechanics of a number of prehistoric technical innovations, have this to say concerning the spear thrower:

From a mechanical point of view its mass should be as small as possible, consistent with the need to maintain its stiffness so that little energy is lost in bending the spearthrower. The stiffest and hence the most efficient spearthrower has a circular cross-section tapering from the hand-hold (where the bending moment has its maximum value) to the hook. Some Australian spearthrowers, such as the Central Australian type, differ considerably from this ideal mechanical form. The Aborigines of the Central Desert had to be highly mobile and were therefore lightly equipped. Their spearthrower was a multi-purpose tool which, in addition to its main function, was also a handle for a stone chisel blade, a shallow container for pigments and blood used during ceremonies, fire-stick, a digging implement, a spear deflector and even a musical instrument. An American curiosity is the bannerstone, a weight that was slipped over the spearthrower [*atlatl* is the name by which the spear thrower or dart thrower is known in the Americas]. The occurrence of these stones has led some archaeologists to believe that the efficiency of a spearthrower can be increased by adding to its mass. Whatever purpose the bannerstone served it had nothing to do with mechanical advantage [because] a heavy spearthrower is inefficient.

The purpose of the stones used on the *atlatl*, or native American spear throwers, was discovered during the 1980s by two engineering students from the Montana State University named W.R. Perkins and P. Leininger.

Although Cotterell and Kamminga were right to say that the added stone weight would not have conferred any mechanical advantage simply because it would have made the spear thrower heavier, it seems its mechanical advantage lay elsewhere, that is to say in its role in co-ordinating the release of energy from the *atlatl* and from the projectile itself. The reason that the stone was 'slipped over' the spear thrower rather than fixed on to it in some more permanent fashion was so that the position of the stone could be adjusted in order to best co-ordinate the energies of the spear thrower and the spear on any given occasion. Perkins and Leininger also realised that the mechanical principles involved in the use of the *atlatl* were comparable to those of the bow and arrow.

With the present state of knowledge concerning these prehistoric weapons, it is impossible to say whether the bow and arrow system developed from the spear thrower or not. Whilst spear throwers made from antler are known from the Magdalenian period, it is possible that the device existed much earlier, probably in wooden form, and that such highly perishable artefacts have simply not survived for archaeologists to find. Some have claimed that its use may go back to the early Upper Palaeolithic. The bow and arrow is known to be at least as early as the Mesolithic period, but many prehistorians are reluctant to call the small projectile points of the Upper Palaeolithic period 'arrows' without the corroborating evidence of the remains of a bow or at least the arrow shafts. There is some supporting evidence for the use of the bow and arrow from the early Upper Palaeolithic period. Experiments using replicas of Aurignacian projectile points found in the Levant showed that when they were shot as arrowheads they broke on impact in the same way that the originals had. In other words, the break patterns and damage of the spent replica arrowheads were of the same form as the used Aurignacian projectile points were found in, therefore implying that the prehistoric projectiles were in fact shot from a bow. Without knowing the relative antiquity of the spear thrower and the bow and arrow, it is impossible to see the mechanics of one evolving from the mechanical principles of the other. Therefore this early chapter in the history of mechanics still remains something of a mystery.

Although the artefacts described in this chapter are essentially practical in their function, they have, both in the prehistoric past and in more modern times, also had symbolic connotations that have raised them above the level of the mundane. The Magdalenian spear thrower with the curious but nevertheless beautifully carved depiction of the defecating ibex clearly points to the considerable artistic investment in this object. It is hard to imagine that this spear thrower was not greatly valued by its owner, and the symbolism, bizarre and enigmatic as it may be, must surely indicate the cultural importance of this early mechanical innovation in Magdalenian times. In many parts of the world the various types of fire-drills have been seen to be symbolically related to the sexual act, with the drill being a widely used metaphor for the penis and the hearth symbolising the vagina, their conjunction engendering the fires of sexual passion.

Among the Aborigines of the eastern part of the Western Desert of Australia, the different kinds of artefacts that made up their minimalist technology had very clear roles in the defining of the traditional division of labour between the sexes. Most of the artefacts were made of either stone or wood. Among the stone artefacts grindstones used for processing edible grass seeds were invariably associated with women, whilst the small grindstones used to prepare the pigment ochre and tobacco were used by both men and women. Artefacts made from flakes of stone which were often hafted in the form of an adze were associated with men, whilst, on the other hand, simpler stone tools in the form of choppers or hand-axes were often used by both sexes.

In other parts of the Western Desert women were forbidden to make use of specific kinds of lithic resources such as flint and chert, which were seen as a male prerogative. Among the wooden artefacts, the hunting spear and the spear thrower were male, and not only were they made exclusively by men but females were not even permitted to touch them; even the places where the kind of wood used to make them grew were strictly out of bounds. The digging-stick and wooden bowls were the main wooden artefacts associated with women. The digging-stick had a number of uses (including extracting animal and plant food from the ground), as did the bowls, which acted as multi-purpose containers for holding food, transporting water or even carrying babies. The anthropologist Annette Hamilton has made a detailed study of the social importance of technology in this region and suggests that:

> The technological apparatus and skills used by women for the manufacture of their wooden implements is a continuation of the older 'core tool and scraper' tradition [of stone tool technology] which appears archaeologically prior to about 4,000–3,000 BC in many parts of Australia. The spear thrower, with its associated adze-stone, perhaps represents a more recent innovation, one which was not made available to the women. It seems likely, for a number of reasons ... that technological innovations in lithic industries adhered solely among men. Women continued the older traditions in technology, as ... they continued the older ritual traditions, not because they are innately 'conservative' but because innovations in both areas are introduced and elaborated within the context of exclusively *male* rituals.

The larger grindstones used by women to make seed cakes were usually too heavy to be transported from place to place and they were thus left at specific locations which were returned to periodically. The grinding of seeds was such an important part of their work in the Western Desert that a whole host of myths and songs were told and sung by the women of the region.

However, unlike all their other daily tasks, which they considered pleasurable pastimes, the grinding process was not enjoyed at all; for them it was truly a grind. It seems that this aversion to grinding was not simply due to the fact that the actual process was tedious and hard, but also because the surplus food that they were able to produce by this means was not something they themselves were to benefit from in any significant way. For the making of large numbers of seed cakes allowed men the freedom to periodically desist from intensive hunting and rely on the food produced by the women whilst conducting large-scale ceremonies which went on for days or even weeks and were only attended by men.

Bearing in mind the 30,000-year-old history of the grinding of seeds in Australia, one can but wonder whether this appropriation of the fruits of women's labour by man that is known from recent times was also the case in remote prehistory, and whether the rich ceremonial life of the Australian Aborigines was built on the foundation of the grindstone. We can see that even some of the early aspects of prehistoric technology, such as the spear thrower, the grindstone, the fire-drill and even the simplest of stone tools, were not simply inanimate, passive artefacts, but were objects that played active social roles and were invested with symbolic attributes which we can rarely recover from the inscrutable nature of the archaeological record.

Chapter 12

The Stone Age Mining Industry

Although mining is nowadays usually associated with the extraction of coal or precious metals from the earth, the earliest known prehistoric mines were dug to obtain different raw materials. The mining industry during the Stone Age was, in large part, concerned with obtaining high-quality flint that had not suffered the adverse effects of weathering that made flint found on the surface an inferior raw material for the manufacturing of stone tools. The mining of flint did not altogether disappear with the widespread switch to metals that took place in later prehistory. Even into our own era flint was still extracted from the ground in order to manufacture gun-flints. In Britain the most famous relic of this very ancient tradition was the flint-knapping industry at Brandon, East Anglia, which continued well into the twentieth century. Its heyday seems to have been in the early nineteenth century when the sheer size of the dumps of the waste products from the industry that had accumulated by the end of the Napoleonic War was commented on by visitors to the area.

In 1879 Sydney B.J. Skertchly published a special report on the Brandon industry entitled *On the Manufacture of Gun-Flints*. In this work he suggested that the industry could be, in his opinion, a direct if very distant continuation of Neolithic flint mining that was already known in his time from the discovery of Britain's major prehistoric flint mine – Grimes Graves, nearby in the county of Norfolk. Skertchly's suggestion turned out to be far-fetched, but not so far-fetched as the name that had been given to the site of this Neolithic mine by the locals. The name Grimes Graves probably derives from an epithet of the Devil (Grim), and so means something like 'Devil's Holes' or 'The Diggings of the Devil'. This name was originally given to the site by the Anglo-Saxons, who named many English earthworks whose origin was unknown to them Grim or Grime. Whilst it had, no doubt, long been thought to be the workings of someone other than the devil, it was not until 1852 that modern investigations into Grimes Graves began. In this year the Revd S.T. Pettigrew took an interest in the devil's work and opened up two of the pits. In the next few years several other clergymen investigated the site, including the Revd Francis Bloomfield, Revd C.R. Manning and Revd William Greenwell. Since these early probings of the site, a number of excavations have taken place, the most extensive being a large-scale project undertaken by the British Museum over a number of years in the 1970s.

The area of the site is about 37 hectares and has an unusual appearance, on account of more than 360 saucer-like depressions that mark its surface. These depressions are the sites of prehistoric mine shafts that were subsequently filled in. That there were over 360 shafts makes it abundantly clear that this was a large-scale operation. Needless to say, the area is rich in flint and the three seams are of different quality. The uppermost (known as 'toppings' or topstone) consists of nodules (lumps) of flint that, due to their being badly weathered, were of the lowest quality and of little value to the flint knapper. The middle layer (wallstone) was also deemed of inferior quality. About 2.5 metres below the wallstone was the real goal of the miners, the floorstone, unweathered tabular nodules of black flint of ideal quality for manufacturing tools. In order to get at the flint, the Neolithic miners not only dug shafts down into the earth but also made horizontal tunnels off the bottom of such shafts, creating what are known as galleries in order to be able to extract a greater amount of flint from the seams (see *Plate XVII*). The making of bell pits was another method that was used in a number of prehistoric mining operations, and was a development of the simple digging of trenches to open up a flint seam. The making of a bell pit involved undercutting at the base of the pit in order to extract as much of the floorstone as possible without causing a cave-in (see *Figure 27*). Mining activity seems to have first started at Grimes Graves in the late Neolithic period (at about 1800 BC), and flint continued to be extracted into the Early Bronze Age. After this time the mines were seemingly abandoned, although the area was reoccupied later in the Bronze Age, perhaps partly on account of the huge amounts of flint lying on the surface, left by the earlier miners, and useful for making tools. In 1971–2, Roger Mercer, an Inspector of Ancient Monuments, supervised the excavation of one of the mine shafts at Grimes Graves. After the excavation Mercer stated that:

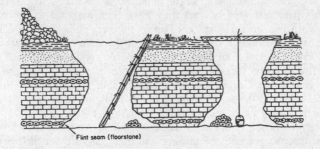

Flint seam (floorstone)

Figure 27

Experience gained during the excavation and available from experimental chalk digging exercises ... elsewhere enabled some approximate calculation to be made of the work involved in the digging of the shaft. Of course with a shaft of limited size there must be fairly close limits to the number of miners who could be involved in its digging. Six or seven men would appear, on grounds of experience, to be the maximum number who could work effectively in the body of the shaft. Taking this figure as a basis for calculation, such a work force would require a further six or seven persons to manage the evacuation by porterage of the produced spoil. This total workforce would take thirty-two consecutive working days to dig the 1971 shaft at Grimes Graves. The working of the galleries would add a further thirteen days to the task. After the evacuation of 800–1,000 tonnes of chalk and sand over the period indicated above, 8 tonnes of flint would have been produced – all of it at the very termination of the exercise. The broken-up blocks of flint once out of the shaft were worked into a variety of tools, notably discoid knives together with some axes.

This considerable amount of high-quality excavated flint was clearly in excess of the quantity required by the miners for their own use, and trading networks for the distribution of this prize flint were most likely to have existed in Neolithic times. The scale of the mining operation at Grimes Graves suggests that some specialisation of labour was involved. The main tool of the miners was the red deer antler pick with which they dug out the flint (see *Figure 28*). It is also very likely that they used wooden ladders, leather bags or baskets to remove rubble and flint, and a number of other tools such as shovels. Detailed examination of the antler picks (which have been preserved in vast numbers at Grimes Graves; one estimate puts the total number used at the site at 40,000) has shown that the great majority of them are made from cast antlers, which are the hardest and most durable kind. The number and quality of the antler picks has led Mercer to suggest that this particular type of tool may have been supplied to the mining community by a service industry set up for this very purpose. Juliet Clutton-Brock, who undertook a special analysis of the antler picks as part of the 1970s archaeological team, compared the selection and preparation of antler picks from Grimes Graves with another site in southern England. The superiority of the Grimes Graves picks in both the quality of the raw material used and the actual making of the tools indicated to her that the miners there may have 'constituted an industrial élite'.

Indigenous peoples from as far afield as Bolivia, Australia and Melanesia believe that the practices of mining and quarrying are watched over by spirits and deities. In some parts of the New Guinea Highlands the intimacy of stone extraction and spirituality is epitomised in the belief that the raw material from which axes and adzes are made is both born and able to reproduce itself in the rocks from which it was taken. Controversial finds at

Figure 28

Grimes Graves indicate that the mining that took place there may have been accompanied by rituals, perhaps to propitiate the spiritual owners of the flint seams and to gain protection from the very real dangers that haunted life underground. Almost 400 objects fashioned from chalk have been found over the years at Grimes Graves, but most are of an unremarkable nature. However, a few ritual objects made of chalk were found during the 1930s excavation led by Leslie Armstrong. Gillian Varndell of the 1970s archaeological team that excavated at the site managed to find his only written note concerning the actual find:

The Grimes Graves figurine [see *Plate XVI*] of the Mother Goddess was found in [July] 1939 during the excavation of Pit XV, shortly before the bottom of the shaft was reached, though the heads of the eight galleries were all exposed. The Goddess was uncovered on the SW side of the pit, at the right-hand side of Gallery 7, in an upright position resting upon a flat slab of chalk which eventually proved to be the top of a pedestal built of similar slabs. Removal of the remaining filling of the shaft revealed upon the floor of the pit, opposite to the Goddess, an ogive shaped platform [i.e. shaped like a pointed arch], believed to be an altar, constructed of blocks of mined flint packed closely together with its apex pointing towards the Goddess. Lying upon it were seven deer antler picks. The altar was excavated with extreme care to avoid disturbance of the flint blocks or the picks, all of which were left *in situ*. At the base of the altar and resting upon the floor of the pit was a skilfully made chalk vessel 3 inches high and 3 inches wide, which had originally possessed a short cylindrical handle, but broken off by the infilling of the pit. Near the NE wall of the shaft & immediately opposite to the Goddess, in line with the centre of the altar, traces of wood ash & charcoal indicated the site of a small fire which had probably played some part in the ritual performed there. Finally, upon excavating the blocked entrance of the gallery which upon its right side was flanked by the pedestal & figurine of the Goddess, a sculpted chalk phallus was found upon the floor close to the left-hand wall of the gallery; also in the central area, a group of three selected small natural nodules of flint, arranged in the form of a phallus, and, near by, a larger nodule of ovoid shape.

The authenticity of this goddess figure and its attendant artefacts has been questioned ever since its discovery; for some observers, the near-perfect alignment of these prehistoric ritual objects is literally too good to be true. It has been suggested that Armstrong himself was the forger, but if the object was not genuine then it is more likely that he himself was the victim of a cruel hoax. Matters were not made any clearer in 1983 when Kevin Leahy of Scunthorpe Museum met 91-year-old Mrs Ethel Rudkin, who had been both a good friend of and a fellow excavator with Armstrong. She had asked Leahy to come to her house as she had decided to donate some objects to his museum. It transpired that she had kept under her bed a copy in chalk of the goddess figurine. She explained to Leahy the story surrounding this copy. She had been at Grimes Graves in 1939 and described Armstrong's behaviour as unusual, as he would not allow her to witness the ongoing excavations. On the day of the discovery she was waiting in her car when Armstrong produced the goddess and the chalk phallus, asking her to look after them whilst he returned to the pit. To while away the time whilst she was waiting for Armstrong to return, she decided playfully to imitate the goddess figure and fashioned an approximate likeness out of a nearby piece of chalk; she then placed the two goddesses together on the car seat. Armstrong was not only not amused; he was furious, and the row that ensued marked the end of their relationship. Mrs Rudkin's frivolous fashioning of the imitation goddess was clearly not meant to offend her friend. Was his fury fuelled by the belief that her actions had trivialised his find? Or did he feel that she was indirectly casting doubt on the authenticity of the goddess by showing how easy it was to make such a figurine out of the local chalk?

Mrs Rudkin made it more than clear to Leahy that she believed that the original figurine shown to her by Armstrong was absolutely authentic. The diaries of both Armstrong and Mrs Rudkin show unusual gaps in the month of July, and this may have more to do with the upset caused by their falling out than the desire to hide any details of the events that transpired at the time of the find. The discovery of a finely made chalk sphinx's head in the Armstrong collection – made by either Armstrong himself, which seems unlikely, or by a third party – shows that Mrs Rudkin was not the only one able to produce such objects for their own or others' amusement. Nevertheless, the existence of such objects does not prove that the goddess and the other ritual artefacts were forgeries. Yet without techniques to prove or disprove their authenticity, the case remains open. The goddess, who was found at the entrance of a prehistoric mining gallery, now resides just inside the entrance to the prehistoric gallery in the British Museum, where she is still exhibited as a genuine object.

The flint mines of Spiennes, just outside the city of Mons in Belgium, were first revealed at about the same time as Grimes Graves. In 1860 it was recognised that the site was rich in Neolithic flint artefacts, and some six years later it was realised that it was in fact a mining complex. Subsequent excavations revealed the full extent of the industrial activity that had taken place in prehistory. The mining industry at Spiennes flourished between 5,000 and 4,000 years ago, and thus was earlier than that at Grimes Graves. As at Grimes Graves, the central motivation of the miners was, of course, to get access to seams of superior flint. R. Shepherd, who in addition to making a detailed survey of prehistoric mining techniques, has working knowledge of the modern industry, has commented that, particularly at Spiennes, the miners' clear grasp of the basic principles of geological stratification is abundantly clear. The initial archaeological excavations at Spiennes in the nineteenth century showed that the miners had dug shafts through five seams of flint in order to work a sixth. Further excavations during the period 1912–14 conducted by de Loë led to the discovery of shafts as deep as 15 and 16 metres. These shafts crossed twelve seams of flint as the miners dug their way down to the thirteenth seam from the surface! De Loë was convinced that Spiennes was a specialist mining community able, as we have seen was most probably the case with Grimes Graves, to export its exceptional-quality flint over long distances by means of river transport.

Clearly the discoveries made at both these important sites show that mining was by no means in its infancy during the Neolithic period – the existence of specialist mining communities (perhaps supported by subsidiary service industries), the trading networks in high-quality flint, and above all the knowledge of geology, extraction techniques and mining practices (the digging of deep shafts and of galleries) all indicate that the industry was, by this time, already highly developed. Günter Smolla has made the point that there are also technical similarities with other comparable activities, such as the digging of wells, which is known from the Neolithic period in both Europe and China. Following this line of thinking, he proposes that flint mining, well sinking and other such technologies were intimately associated with the rise of what was once called the Neolithic Revolution.

Many archaeologists have sought to explain such sophisticated flint mines as those described above as a development that was adopted from technology originally used in extracting copper. Although the quarrying and use of flint preceded copper mining, mines such as Spiennes and Grimes Graves which involve developed mining techniques (that is, the digging of shafts and galleries, etc., rather than just quarrying) are much later than the copper mines of the Balkans, which date back to the fifth millennium BC. It is for this reason that some scholars have sought to derive genuine flint mining from 'higher', more 'advanced' cultures that had established traditions of copper mining. The copper mine at Rudna Glava, 140 kilometres east of Belgrade, is at least 7,000 years old, according to Marija Gimbutas, and has vertical 'shafts' as deep as 20 metres; antler picks were also excavated at the site. The excavation in 1972 by E. N. Chernykh of the copper mine at Ai-bunar in central Bulgaria (dated to the middle of the fifth millennium BC) revealed 'shafts' over 100 metres long and up to 20 metres deep. Clearly the scale of these mines is impressive, but strong criticisms of the idea of seeing them as technological prototypes for the later flint mines of England and Belgium have been raised by Gerd Weisgerber. He is dissatisfied with the use of the term 'shaft' in describing the extraction industries of these and other copper mines. Concerning Rudna Glava he has this to say:

Where the [copper] ore outcropped on the surface, it was simply dug out in slanting pits irregularly sited over the area. The 'shafts' simply followed the steeply inclined veins of ore down to ground water level. In mining terminology, such works would not be termed 'shafts', as the inclined tunnels were not constructed to provide access to underground galleries.

On these grounds he rejects the possibility of such copper mining being comparable in technology to developed mines such as that of Spiennes. He is equally dismissive of the 'shafts' at the Bulgarian site of Ai-bunar, which he views as more of a large-scale quarrying project than developed mining. In reviewing the technology of mining in prehistoric Europe, he concludes that techniques used in copper extraction did not influence the development of flint mining, since the latter was more sophisticated. In fact, he states that the copper mining technology only reached a comparable level around 1200 BC in the Late Bronze Age!

Having dismissed attempts to see underground flint mining as a spin-off of copper mining, we can return to the question of developed mining as an integral element of Neolithic innovations. At present no underground mining is known to have taken place in Europe. There is evidence from a number of Palaeolithic sites in Poland, Hungary and Switzerland that minerals were being extracted from the ground (the Löwenburg site in Switzerland which provides evidence of an open-air mining operation and red deer antler 'picks' is attributed to the activities of Neanderthals), but none of these instances involve underground mining, and they are better described as quarrying and simple mining from the surface. According to Weisgerber and others, mining is not known in Europe earlier than 4000 BC. The evidence from Europe seems to suggest that Palaeolithic extraction technology was restricted to quarrying and that systematic mining was an innovation of the Neolithic period. But in archaeology things can change dramatically overnight, and that is exactly what happened in this case. New evidence from Palaeolithic sites in Egypt discovered in 1987 put underground mining further back in the depths of time; nearly 30,000 years earlier than had previously been known.

In their report of these discoveries, Pierre Vermeersch, Etienne Paulissen and Philip Van Peer describe two sites (Nazlet Khater 4 and 7) in the Nile Valley that provide clear evidence for underground chert mining during the period 35,000 to 30,000 BP. The main excavations took place at Nazlet 4, and most of the information consequently comes from this site. The Upper Palaeolithic miners dug vertical shafts down to a depth of two metres, and some were extended at the bottom to make bell pits. Ditches were also dug down to the same depth and underground galleries were discovered extending from both the ditch sides and the base of the bell pits. The largest of the galleries investigated by the archaeological team had an area of approximately 10 square metres. In places they were able to determine that some of the galleries had collapsed. Mining tools were also found within the galleries and are of two basic types: rough hammerstones and picks. The picks are reminiscent of the Neolithic examples found at Spiennes and at Grimes Graves, but in the Egyptian case they are made of hartebeest and gazelle horns. Nearby, and to the north-west of Nazlet 4, they discovered the site of Nazlet 7, which was also an underground mine.

Even earlier evidence for the systematic extraction of chert was found by the team, dating back to the Middle Palaeolithic period. Quarrying activities at the sites of Nazlet Safaha 1 and 2 consisted of the digging of ditches and pits and have been dated to the period after 60,000 BP (probably between 50,000 and 40,000 years ago). Middle Palaeolithic pits were also found at another site (Taramsa 1), which, judging from the diverse kinds of stone tools discovered there, seems to have been frequented by numerous groups of people during this period. No digging or quarrying tools were found at any of these Middle Palaeolithic chert extraction sites, which may indicate that such implements were not used or that they were made of organic materials (e.g. wood or horn) that have not survived in the archaeological record. Both the Middle and Upper Palaeolithic Egyptian sites provide us with important information for the reconstruction of the prehistoric development of mining. That systematic quarrying existed in the Middle Palaeolithic and underground mining in the early Upper Palaeolithic means that previous evolutionary models have to be drastically revised. We have seen how previously systematic quarrying was seen as the extent of the Upper Palaeolithic ability to extract raw materials, and how underground mining was perceived to be a Neolithic development. Now that whole sequence has had to be shifted back in dramatic fashion.

Chapter 13

Ochre: Blood of the Earth

Ochre is the name commonly given to the iron ores haematite, goethite and limonite which have long been used as a source of red, yellow and brown pigments respectively. Of these it is haematite that figures most prominently in the archaeological record, and most references to ochre refer to this specific ore. Whilst the metallurgical use of iron was a comparatively late development in the prehistory of technology, and is of course known as the Iron Age, the use of ochre goes back to the Lower Palaeolithic period. In telling the story of ochre it is necessary to transcend the technological sphere, for this material has been implicated in the origins of symbolic behaviour, of art, and of religion itself. The famous archaeologist and palaeoanthropologist Louis Leakey, whilst excavating at the site known as BK, Upper Bed II (dated to approximately 1.2 million years ago) at Olduvai Gorge in Tanzania, came across two lumps which he thought were ochre. Subsequent analysis showed that they were actually volcanic ash. Thus the story does not begin that early, and it is nearly a million years before we come across the earliest instances of ochre use.

Interestingly enough, the oldest site indicating the use of ochre also provides the earliest evidence of possible mining activities. Preliminary dating of the Wonderwerk Cave in South Africa suggests that ochre mining at the site may date back to 350,000 or even 400,000 years ago, according to Peter Beaumont. Lying among the hand-axes found at the site are numerous pieces of ochre that seem to have been chipped off local rock. At the 300,000-year-old site of Terra Amata in France, a number of pieces of ochre were found in association with Acheulian tools. Inspection of the lumps of ochre revealed the unmistakable signs of wear marks that pointed to the fact that *Homo erectus* had used them for an unknown purpose. A similarly marked piece of red ochre was discovered by J. Fridich at the 250,000-year-old Acheulian rock shelter site of Beçov in Czechoslovakia. He also found a flat stone that had been used to prepare ochre powder. Marshack describes the unusual way in which Fridrich was able to reconstruct an archaic event that had taken place at the shelter:

> On the floor of the shelter, at the side where the piece of ochre was found, there was a wide area of ochre powder. Seating himself on a rock against the wall of the shelter to study the ochre, Fridrich found that his feet

accidentally fitted the only two areas without ochre powder. *Homo erectus* had sat on this stone, away from other activities in the site, while he made his red powder.

Archaeologists have found the desire to speculate on what uses *Homo erectus* may have had for ochre impossible to resist. Those who are inclined to ascribe such hominids with the capacity for symbolic behaviour have suggested that finds such as these may indicate the practice of body-painting, and the red colour of ochre may have acted as a symbol of blood. On the other hand, those who see the hominids of this period as little more than tool-using animals whose activities were confined to the utilitarian sphere suggest that ochre may have been used in the treating of animal skins and hides in order to make some rudimentary form of clothing or bedding. They also note that the application of ochre to the human body does not always indicate symbolic body painting. In fact there are examples of tribal peoples who rub ochre on their skin in order to repel insects and provide protection from the sun.

If we leave the prehistoric evidence for a moment, we find that a brief perusal of ethnographic accounts of the applying of ochre to the body do not make matters any clearer. An example of a strictly practical use of ochre is that of the Himba people of Africa, who mix grease and ochre together and rub the mixture on their bodies in order to protect themselves against the sun and insects. The native inhabitants of the isolated Andaman Islands in the Bay of Bengal are known to have smeared newly born babies with ochre as a form of 'protection', which indicates a mainly practical function but with some magical elements. The Gugadja people of north-western Australia used ochre as a medicine; after having moistened it with water or saliva, it was applied to sores and burns and was also used to treat internal pains. The medical properties of ochre are attested to by both Western and Eastern traditions, and its value is largely due to its antiseptic qualities and its ability to staunch bleeding. The use of ochre in burial rites and as a painting material is common in the tribal world, but as both these practices are well attested in the later stages of the Stone Age, there is no need to describe any of the more recent examples. Even from the few cases of its modern use mentioned here, it is obvious that ochre has a number of practical and symbolic uses and that these two spheres of activity are often impossible to distinguish. Ochre use can be both practical and symbolic at the same time, and this may well also have been the case in early prehistory.

Discoveries of the Lower Palaeolithic use of ochre are not confined to this single site nor to the continent of Europe. Nodules of ochre have been found at the Acheulian site of Hunsgi in southern India, and have been dated to the period between 300,000 and 200,000 years ago. They were excavated by K. Paddayya. Robert Bednarik, who has examined the pieces, found one of them particularly striking:

One of the ochre pebbles, measuring only about 20 mm and slightly tapering at one end, immediately caught my attention. It bears a 7 or 8mm-long facet extending from the vaguely pointed end which is entirely covered by striations [i.e. marks]. The marks are just barely visible to the unaided eye, but under magnification they provide important information ... the marks are extraordinarily well preserved. Their distinct restriction to the one facet seems to rule out the possibility that they are the result of sedimentary movement ... Moreover, the abrasion occurred in several successive episodes ... which renders a natural cause extremely unlikely.

Having established that the marks were very likely to be artificial rather than natural, Bednarik suggests that pieces of ochre such as this one may have been used as a kind of crayon to make markings on rock. In short he sees them as possible evidence pointing towards an Acheulian tradition of art existing in India at this time. After the Acheulian period, the number of ochre finds increases quite markedly. We should be cautious in simply attributing this to an increased interest in and need for ochre by later Stone Age peoples; it is certainly due, at least in part, to the fact that the further back one goes the less likely it is that evidence for this (and many other) activities will have survived.

There have been several significant finds of ochre from the Middle Stone Age period in Africa. Ochre 'crayons' have been found at a number of sites, in some cases along with stone slabs with patches of ochre staining. The most interesting of these is what is generally acknowledged as the world's earliest known mine, at the Lion Cavern in Swaziland. The dating of the site is rather problematic, but it is certainly no later than 42,000 years old and the evidence supplied by the tools (including actual mining tools) found there has led Peter Beaumont to the conclusion that mining for ochre pigments began about 120,000 BP. Thousands of stone tools were found at the site and indicate that the mine was in use for many thousands of years. Estimates suggest that more than 1,200 tonnes of ochre had been extracted during the lifetime of the mine by cutting a large horizontal gallery into the cliff face. Another important discovery from Swaziland is the burial of an infant at Border Cave. The almost complete skeleton of the infant, who was between four and six months old at death, was buried in a shallow grave. Some of the bones were stained with ochre and the skeleton was accompanied by a perforated shell that must have originated from the seashore over 80 kilometres distant. The burial may be as old as 100,000 years according to some estimates, but it seems far more likely to be less than 40,000 BP. The discovery of this ochre-stained skeleton does not necessarily point to the ritual use of ochre in burial ritual, as it is possible that the body was wrapped in a hide or skin that was tanned by ochre. As I have already mentioned in Chapter 11, a number of grinding stones that still retain traces of the ochre which they were used to process have been found at the Bushman Rock shelter site in the eastern Transvaal and have been dated to about 43,000 to 47,000 BP.

In the Mousterian levels of the Pech de l'Azé cave site in France, François Bordes discovered abundant signs of mineral pigments used by Neanderthals. In addition to a few fragments of red ochre and traces of yellow ochre, he found more than a hundred pieces of manganese. The manganese produces a black pigment which was more common in the environs of the cave than ochre. But we cannot conclude that black was the Neanderthals' favourite colour simply because there was so much more of it in use at the site! As is the case with ochre at other sites, the manganese showed signs of having been deliberately shaped in order to make 'crayons'. In 1908 a Neanderthal burial was discovered at La Chapelle-aux-Saints in south-west France; alongside the skeletal remains of this old Neanderthal man a number of lumps of ochre were found, and it has been thought that these might have been grave goods, or offerings for the afterlife. The discovery at the Mousterian site of Tata in Hungary of a mammoth-tooth plaque that had been covered in red ochre points to another possible symbolic role for this mineral substance in the Middle Palaeolithic period.

Ochre was also in demand from the early Upper Palaeolithic period onwards and, as in Africa, was mined. The excavation of an ochre mine near Lovas in Hungary (dated between 40,000 and 30,000 BP) in the early 1950s was interpreted by its excavators (Mészáros and Vertes) as a concerted effort to extract the material for use as paint. A large number of implements, including those used specifically in the digging for ochre, were found at the site – bone scooping-tools, a boar's tusk digging tool, stone chisels, antler pick-heads, bone awls and an 'ornamental' bone tool fragment incised with parallel lines and notches. Although the excavation report describes it as ornamental, it may have been a tally of the sort described in earlier chapters (i.e. some kind of artificial memory system). Containers made from hollowed-out antler tines were used for storing paint, whereas at other European sites of the Upper Palaeolithic bird bones fulfilled the same purpose. The paint mine at Lovas is identified with the Szeletian culture on account of the distinctive leaf-shaped stone tool found there which is the trademark of this culture. This culture grew out of the preceding Mousterian tradition and bridges the Middle and Upper Palaeolithic divide. The excavators of this ochre mine were clearly surprised at what they had found:

> The Lovas find, in its unity as a harmonious whole, appears unexpect-edly, out of nothing, as it were. The stock of implements found at other palaeolithic sites, of the same age and culture as Lovas, serves almost exclusively the fundamental aims of self-preservation: it consists of articles of hunting and clothing and the tools used to produce these articles. As against this, the Lovas find consisted of tools suited to quarry red paint – a purely 'luxury' article, according to our present outlook. The quantity and perfect finish of the tools, together with the difficulties

involved in obtaining the raw material, demanded an astonishing degree of concentration and exertion of will, tending to a special object, on the part of primitive man. Such qualities are usually not associated with palaeolithic man who is regarded as a being unable to concentrate his attention, rather clumsy and heavy in his cerebral activities except those connected with the fundamental functions of self-preservation and the propagation of the race.

The discovery of this site shows that in Stone Age Europe mining for ochre was probably even earlier than the systematic extraction of flint; the use of antler picks at Lovas is distantly echoed by comparable tools used in the Neolithic mines. It is also possible that the ochre from Lovas was traded in a similar way to the flint from Spiennes and Grimes Graves, although probably over shorter distances. The diggings at Lovas did not involve subterranean mining of the sort that was being undertaken in the Nile Valley at a similar date, but the Hungarian ochre mine nevertheless shows that from the very start of the Upper Palaeolithic – and perhaps somewhat before the start – mining was very much part of cultural life. Another aspect of the importance of ochre use in early technology is revealed by another Palaeolithic site – Arcy-sur-Cure near Paris in France. This site belongs to the transitional phase between the Middle and Upper Palaeolithic in France, to the period known as the Chatelperronian (also called the Lower Perigordian), and Neanderthal remains have been found at the site. Ochre was prepared for use at this site on a large scale; the excavator Leroi-Gourhan found large patches of ochre on what appears to have been the floor of a hut measuring three or four metres across. It is unclear whether the ochre was being prepared for some practical purpose such as the tanning of hides, or whether it had some magical or aesthetic value for those who were using it.

Denise Schmandt-Besserat has stressed further details concerning this site that are vitally important in reconstructing the development of technology. Grinding stones and pestles found at Arcy-sur-Cure indicate that these two types of artefacts which are usually identified with much later developments that heralded the dawn of agriculture (from their role in the processing of grain) were initially used in the preparation not of cereals but of ochre. She also noted that the ochre found in hearths at the site was not likely to have been due to accident or indifference on the part of the people who used it. The calcination of ochre produces a wider range of colours than can be obtained from ochre that has not been subjected to heat treatment. The temperature required to achieve this (approximately 260°–280°C) is easily achieved in a hearth fire. This evidence for the use of heat treatment intentionally to alter ochre has implications for the origins of heat treatment in both lithic and ceramic technologies. The earliest known lithic heat treatment from the Solutrean period as well as the first evidence of ceramics at the Gravettian site of Dolni Vestonice are later than the thermal alteration of ochre by the Chatelperronians. Based on our present knowledge, it seems that the pyrotechnic arts were first applied to ochre and that this technology may then have been transferred to other materials such as flint and clay.

From the Acheulian period down to the early Upper Palaeolithic, there is clearly considerable evidence that ochre was being used, but in most instances we do not know what it was being used for. There is no reason to think that its use was confined to a single purpose; a single group of ochre users could have made many uses of it – as a medicine, a paint, a hide-tanning agent, to protect the skin and in the performing of rituals. The use of ochre continued throughout the rest of the Upper Palaeolithic, where its most visible role was in providing pigments for use in painting. Recent analytical techniques applied to the study of Magdalenian cave paintings in France have shed new light on the making of paints used by prehistoric artists. Jean Clottes, Philippe Walter and Michel Menu took nearly 60 minute samples of red and black paint from the Palaeolithic paintings that adorn the Niaux cave in the French Pyrenees in order to analyse their composition. The results 'clearly demonstrate that the Niaux artists created truly complex paints by mixing their pigments with mineral extenders and binders'. The red pigment was made from ochre and the black was made from manganese dioxide, charcoal or a mixture of the two. All these minerals were locally available, precluding any need to obtain them through trade. The researchers were able to distinguish four kinds of paint recipes which differed in respect of the extenders that had been used.

Such findings are very useful for dating specific paintings, for although the French caves have received enormous attention from scholars during the twentieth century, an accurate chronology of the artwork is still very much in its infancy. Recent advances in scientific means of analysis have allowed prehistorians not only to determine the composition of pigments and the natural source of their constituents, but sometimes also to obtain radiocarbon dates from minute samples which can be taken without damaging the paintings. Michel Lorblanchet, a specialist in the painting techniques used in Palaeolithic art, along with his interdisciplinary team, conducted a study of the Cougnac cave in Quercy, France. One of their objectives was to locate the source or sources of the paint pigments used by the prehistoric painters at the site. It turned out that there were two sources: one directly outside the cave; the other a mine about 15 kilometres away. Both these sources consisted of clay-sand sediments containing yellow ochre, whereas the paintings used red not yellow. Experiments were then undertaken in order to see if samples of yellow ochre from both sources could be transformed into red:

> The first stage consists [of] mixing the ochre-coloured sediments with water and then decanting them little by little into a hollow container in order to eliminate most of the sand. The larger quartz grains settle at the bottom, the water containing the clay fraction in suspension is decanted

into a wide shallow dish, where the water is allowed to evaporate. The procedure can be repeated several times, and the amount of decantation determines the proportion of quartz sand remaining in the pigment.

After evaporation the flakes of dried ochre clay are collected and ground. The yellow powder is placed on a stone griddle (preferably a large flat river cobble) and these yellow ochre cakes are then placed over the glowing embers of an open fire of the prehistoric type. After two or three hours of intense heat the yellow ochre turns to a shade of red that is identical to the pigment used by the Palaeolithic artists.

As we have seen, similar thermal alteration of ochre goes back at least to the beginning of the Upper Palaeolithic period. Ochre has been prized since prehistory by the Australian Aborigines as a source of pigment, for they, like the prehistoric inhabitants of France, loved to paint, and indeed they still do. Michel Lorblanchet has spent time in Australia learning much from contemporary Aboriginal artists, who have shown him techniques that he has been able to recognise were once used in Palaeolithic Europe. Of course Europeans still love to paint but in rather different styles than their Stone Age ancestors, but we should not be misled by art critics who tell us we have left this primitive stage in the 'evolution' of art way behind us. Picasso, who of all twentieth-century artists must surely be the one most accredited with 'developing' art, is reported to have said on leaving Lascaux, 'We have invented nothing!'

Ochre crayons found at the Naulabila site (or Lindner site, as it is sometimes known) in Arnhem Land, northern Australia, may be up to 50,000 years old, although this date may be considered way too early. In the same area another site, Malakunanja II, has pieces of red and yellow ochre in the earliest levels of occupation at the site which may be as early as 45,000 BP, although again this date has not received general acceptance. Ochre use at both sites from about 20,000 years ago is widely acknowledged. The most dramatic Australian site relating to ochre is surely Wilgie Mia in Western Australia. The site is a large-scale ochre mining operation that continued to be exploited by the Aboriginal people until comparatively recently. Thousands of tonnes of rock have been removed by Aboriginal miners in their efforts to extract the precious commodity. A hill containing seams of yellow and red ochre has been subjected to major excavation in the course of this mining operation. The cut into the hill ranges from 30 to 15 metres wide and is 20 metres deep. Mining equipment included stone tools, fire-hardened wooden wedges and scaffolding. Traditionally the mine and the activities that took place there were supervised by elders, and the uninitiated were not permitted to enter parts of the mine. There was also a taboo on removing mining tools from the site. The ochre from this and other Aboriginal mines was traded over long distances; that from Wilgie Mia is reported to have reached Queensland. The sacred nature of the landscape and its intimate connection with mythology are well-known cultural traits of the Australian Aborigines, and it is hardly surprising that a mine such as Wilgie Mia should have its own story. According to Josephine Flood's rendition of the tradition:

> The ochre was formed by the death of a great kangaroo, who was speared by the spirit being called Mondong. The kangaroo leapt in his death agony to Wilgie Mia, where the red ochre represents his blood, the yellow his liver, and the green his gall. The last leap took the kangaroo to another hill, called Little Wilgie, which marks his grave.

It is likely that the French cave painting sites were also the subject of mythology in Palaeolithic times, but what we can learn from this aspect of Aboriginal culture is far more indirect than the practical clues that Lorblanchet gained from his Aboriginal teachers. Our knowledge of the ritual activities that must have gone on in the Palaeolithic caves is necessarily hazy and consists only of a few hints left in the archaeological record. These natural cathedrals, distant ancestors of the passage-graves and burial mounds of the Neolithic period, are described tentatively by Lorblanchet, in recognition of our incomplete knowledge of their symbolism and purpose, as 'sanctuaries', 'temples' and 'secret places'. There is every reason to think that analogous artistic works once existed in numerous open-air sites across Europe and beyond, and that these also played a powerful role in Palaeolithic societies. It is not the case that the artists of Palaeolithic France and Spain were inherently superior to those of other regions, but rather that their magnificent work in caves has been fortuitously preserved from the ravages of time whilst the open-air sites have all but disappeared during millennia of weathering. Thus these cave sanctuaries in France and Spain and the artistic treasures they contain represent a mere fraction of the European art of Palaeolithic times, a precious reminder of the imaginative powers of our forebears that once animated the landscape of the continent.

Chapter 14

Venus Figurines:
Sex Objects or Symbols?

Although the cave paintings of France and Spain are the most famous examples of art from the Upper Palaeolithic, there are numerous other works in the form of decorated artefacts, sculptures and figurines that show comparable skill and imaginative powers. Although a greater variety of both paintings and art objects are known from Europe than from other continents, there are numerous works from beyond Europe that are many thousands of years old. Yet, as in the modern art world, not all works of art are genuine. Forgeries of prehistoric objects were common in late-nineteenth-century America, and the primary motivation for the production of such fakes was the desire to 'prove' that humans had existed from early times in the Americas. Whilst most fakes were quickly dismissed, one such piece, the Holly Oak engraving, was only definitively recognised as such a hundred years after it was first announced to the world in 1889 by an American named Hilborne Cresson.

The object in question is a pendant made from a whelk shell and engraved with a depiction of a mammoth, which Cresson claimed to have unearthed at Holly Oak, Delaware, in 1864. Cresson never gave a proper explanation for why he did not announce his dramatic find earlier; after all, to wait 25 years before doing so is strange behaviour indeed. There were other reasons to doubt its authenticity, among them its similarities to a genuine engraving of a mammoth found in the rock shelter of La Madeleine in France in 1864, drawings of which Cresson would have had access to. Cresson himself came to a tragic end, taking his own life in 1894 apparently as a result of anxieties caused by his involvement in counterfeiting money. Although its authenticity was doubted from the beginning, the Holly Oak engraving did have its defenders and Paul Bahn has described the revival and final fall of this tenacious hoax:

> The Holly Oak figure was eventually resurrected from relative obscurity by its supporters, even being placed on the cover of *Science* [in 1976]. Some time later, a correspondence in the same journal revived the debate, with the case for the prosecution being set out at length. Recently, the matter was decided once and for all when the shell was subjected to

radiocarbon dating despite the often inaccurate results of radiocarbon dates from shell, it is pretty obvious that the whelk was taken by Cresson from an archaeological site of around the 9th century AD, and the engraving made on it by him in imitation of the La Madeleine mammoth, probably in the 1880s rather than in 1864 as he had claimed. Ironically, the object will probably attract far more fame and attention as a proven fake than it ever did as a possible antiquity.

There are a number of other objects that have been claimed as evidence of art in the Americas before 10,000 BP, such as a bone engraved with the figure of a rhinoceros found at Jacob's Cave in Missouri. During the last century a carved and engraved bone with nostril holes cut into it was found at Teqiuxquiac in Mexico. It depicts some kind of animal, possibly a dog, and is estimated to be about 12,000 years old. Controversial use of dating techniques on petroglyphs (rock carvings) in parts of North America have produced early dates, a geometric design from Arizona is said to be 18,000 years old and an engraving of an animal in California 14,000 years old. These dates belong to a time period which is earlier than the generally accepted date for the entry of the first people to populate the New World, that is to say, 12,000 years ago. Less controversial are 8,000-year-old incised stone artefacts discovered as far afield as North Carolina and Argentina. Such finds indicate both the antiquity and the widespread distribution of these traditions of rock art.

By far the largest number of such incised stones to have been discovered in the New World comes from the Gatecliff Shelter in Monitor Valley, Nevada, USA, where 428 specimens were found, ranging in age from about 3300 BC to 1300 AD and thus showing a continuous tradition of art that existed for over 4,500 years before contact with Europeans. According to Trudy Thomas, an archaeologist who has made a detailed analysis of these early works of art, the sheer number of engraved stones and their distribution over a wide area show that they represent a major symbolic tradition in the Great Basin area, a tradition that continued for thousands of years. The symbols depicted on these stones are complex arrangements of zigzags and other designs, also known from the rock art of the region. In many cases marks on the stones were superimposed on earlier engravings, indicating the re-use of the material. In other instances Thomas was able to tell from her analysis that not all the markings in a particular sequence were made at the same time, which may perhaps indicate that they were used as some kind of tally. The incised stones from Gatecliff Shelter clearly have parallels elsewhere in the New World and can be dated, as I have said, in both North and South America, to at least 8000 BP. It may well be that the first migrants to the New World brought such artistic and notational traditions with them across the Bering land bridge that once linked the Old and the New World, but with the present state of knowledge this can only be speculation. It is tempting to see certain parallels with the Upper

Palaeolithic tallies of Europe studied by Marshack and others. European notations and tallies are obviously remote from those of the Americas, and without far more evidence for the existence of such artefacts from Asia, the two traditions cannot be shown to have any historical connection.

The discovery of art of Palaeolithic age has been rather recent in many parts of the world, and these new finds are changing the overall picture of the global distribution of such works and showing that the pride of place given to Europe in the early prehistoric story of art may be partly the result of more intensive and extensive archaeological researches on that continent. The first report of Palaeolithic art from China was made as recently as 1991. The object in question is a 13,000-year-old engraved antler fragment found by You Yuzhu at the Upper Palaeolithic site of Longgu Cave in Hebei Province, about 100 kilometres north-east of Beijing. According to Robert Bednarik, who has personally examined the antler:

> The engraved lines were superbly crafted with stone tools, their groove patterning matches that on European and Siberian portable art objects. Line widths and depths, as well as the spacings between the lines, have been carefully maintained throughout. The layout of the three patterns is bold and confident, suggesting the hand of a highly experienced artisan with a repertoire of very distinctive and sophisticated designs. One pattern consists of four sets of six or seven parallel wave lines, competently arranged to form a consistent design; the second is an elaborate figure eight motif; whilst the third is an arrangement of parallel and zigzag lines enclosing two elongate panels of oblique cross hatching. The artist demonstrates an admirable control not only of even spacings and groove depths, but also of the technical aspects of constructing complex curved arrangements and integrating them seemingly effortlessly into a rectangular layout. The definition evident in all three patterns and the distinctiveness of the designs suggest that well-defined meanings were attached to them.

One of the engraved grooves on the antler still retained traces of ochre and it may be that originally all the grooves were coloured red by this means. As Bednarik points out, the complexity of the design is far greater than that known from similar decorated objects from the Upper Palaeolithic period in Europe, and before him Marshack noted that the patterns known from eastern Europe and also in Siberia are comparatively more elaborate than those from western Europe. Although it is dangerous to speculate too much on the basis of this single find from China, it does seem that the use of such symbols was perhaps more developed in some respects in Asia than in Europe during Upper Palaeolithic times. Palaeolithic art is equally rare in other parts of the Far East. Sohn Pow-Key has suggested that rock carvings in Korea depict extinct animal species and are thus of Palaeolithic age. He has also made the more controversial claim that there is evidence for the

Figure 29

existence of art in Korea during Middle Palaeolithic times in the form of modified bones said to be representations of animals. In Japan a number of engraved pebbles have been found at the Kamikuroiwa site. They belong to the Initial Jomon phase of the site and have been dated to 12,165 BP. Among them are the so-called 'Venus' pebbles, which appear to depict breasts and skirts (see *Figure 29*).

Art with an erotic overtone has been found at the el-Wad Cave in the Mount Carmel region near Haifa, Israel, and belongs to the Natufian culture that flourished at the tail end of the Upper Palaeolithic period from about 12,800 to 10,300 BP. A number of phallic objects made of flint have been recovered at the site. The makers of these objects seem to have deliberately chosen nodules of flint with a shape that required only a few artistic modifications to bring out its phallic qualities. In one particular case the foreskin has been suggested by means of a circular incision towards one end of the object, whilst the urethra was made by drilling a hole in the same end. A remarkable work of erotic art, depicting sexual intercourse and known as

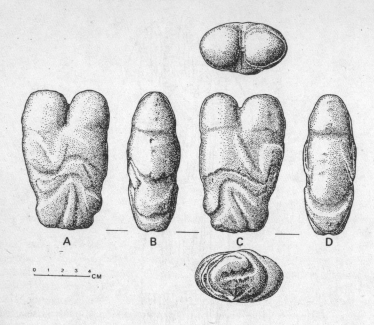

Figure 30

the Ain Sakhri figurine, also comes from the Levant and is probably of Natufian age, although the story of its discovery makes dating it definitively to a particular period impossible.

The object had been found by a local Bedouin in the desert a few kilometres south of Bethlehem and was bought from him and placed in a little museum of prehistoric antiquities set up by French priests in Bethlehem. In 1933 the French vice-consul to Palestine, René Neuville (who was also a prehistorian), and l'Abbé Henri Breuil, the distinguished expert on prehistoric art, viewed the object and were united in their opinion that it was a genuine artefact belonging to the Natufian period. Neuville acquired the object himself and in 1958, a few years after his death, it came up for auction at Sotheby's and was bought by the British Museum, where it is today. Recently Jill Cook at the British Museum and Brian Boyd of Cambridge University have re-investigated the object and its provenance. They concluded that there was no real reason to doubt that the object was indeed very old. However, because it had no archaeological context (i.e. it had not been found by an archaeologist in its original setting) it could belong to either the Natufian period or the subsequent early Neolithic period, possibly even later. Whichever of these periods of the Stone Age it

Figure 31

comes from it is a most remarkable prehistoric depiction of sexual intercourse.

The figurine (see *Figure 30*) is made from a calcite cobble and has been sculpted with the help of a stone chisel to produce a remarkable three-dimensional example of a *double entendre*. For whilst the figurine depicts two lovers in the sexual position shown in *Figure 31*, it also has obvious phallic connotations from every conceivable angle. Viewed from either side A or side C, the sculpture gives the distinct impression of two penises side by side and may perhaps suggest homosexual practices, and in this reading of the piece the heads of the lovers become the heads of the two penises. From sides B and D the phallic effect is also obvious. Looked at from above, the two heads appear to represent testicles whilst the bottom is reminiscent of the end of the glans penis. Both the complexity of sexual symbolism and the brilliance with which the idea was executed in stone make the object entirely unique. By comparison the Natufian phallic objects mentioned above fade into insignificance, although one can see how they may belong to similar if not identical artistic traditions.

Art objects in the Levant that date from before the Natufian era are very . scarce indeed. A slab of limestone incised with an artistic design has been excavated from the Hayonim Cave in western Galilee and dated to the Aurignacian period, about 30,000 BP; that is to say, the early part of the Upper Palaeolithic. The incisions and scrapings on the object were probably made with a flint implement of some kind and the artist seems to have been trying to depict an animal such as a horse or a cow, although it is by no means clear which. Both the rarity of such early finds and their comparative crudity when compared with later works may lead one to assume some kind of evolution from a primitive phase of art in the early Upper Palaeolithic period to a more fully formed tradition in the later Upper Palaeolithic in which both craftsmanship and symbolic expression attained a higher level. Certainly such views were, at one time, very popular when describing the

Upper Palaeolithic art of Europe and beyond. Whilst some of the merits of the art of the early Upper Palaeolithic period were recognised, it was thought that cultural and artistic complexity did not really come about until about 17,000 years ago, at the time of the spectacular paintings of Lascaux.

In the light of recent discoveries and researches, it is now impossible to show that any evolution or artistic progress took place during the Upper Palaeolithic period – it was complex and sophisticated from the beginning to the end. Recent discoveries in the Chauvet Cave in Ardèche, France, have shown that sophisticated cave art such as that at famous sites like Altamira and Lascaux existed at a much earlier date (see *Plate XXIV*). Samples of charcoal taken from drawings in the cave have been reliably dated to more than 30,000 years ago and they are thus *twice as old* as many of the Upper Palaeolithic paintings that were known previously. The idea that Lascaux and other painted caves represent an evolution out of the childish or primitive phase of art of the early Upper Palaeolithic must now be totally rejected.

It had also once been thought that the Upper Palaeolithic peoples who lived between 40,000 and 25,000 years ago had a much simpler social organisation than their counterparts in the later phases of the Upper Palaeolithic, and that this was one of the reasons their art did not flourish to the same extent. But Randall White of New York University has shown that hierarchical societies existed during these earlier times. His study of beads used as body ornaments is a fine example of the type of approach described in Chapter 7 in which apparently minor aspects of a subject can reveal vitally important information that sheds light on the overall picture. Personal ornaments appear in the Aurignacian period in various parts of Europe from about 40,000 years ago. They were made of numerous materials, including fossil shells and coral, animal teeth, ivory, limestone, jet, bone and antler. Yet, as Randall White points out, the selection of materials to be used in making body ornaments was by no means arbitrary or haphazard. Even within one category, that of animal teeth pierced to be hung around the neck, he has been able, by plotting their distribution across the continent, to show regional predilections for the teeth of certain species. The canine teeth of foxes were very much in fashion in Russia, Germany, Belgium and France, whereas these are almost entirely absent from Spain and Italy where red deer canines were favoured. Marine shells used as personal ornaments are much more common in certain inland regions, in which their exotic origins seem to have added to their desirability.

At the Upper Palaeolithic site of Sungir near the Russian city of Vladimir, a number of 25,000-year-old burials were found, and three of the bodies were elaborately embellished with thousands of beads made from ivory and other personal ornaments. The first burial was that of a 60-year-old man whose adornments consisted of 2,936 beads and fragments that had been arranged in strands on various parts of his body (see *Plate XXII*). Judging from the arrangement of beads found when his body was excavated, it seems that he wore a beaded cap that also was decorated with a number of fox teeth. Around his neck he wore a flat pendant that had been painted red with a black dot on one side. On his arms were a total of 25 thin ivory bracelets made from mammoth tusk, each of which was pierced at both ends to allow them to be fastened in circular form.

The second burial, identified by Russian physical anthropologists as that of a 13-year-old boy, was even more remarkable: 4,903 beads of the same kind as those found on the man (although of smaller size) were similarly arranged in strands, and he also wore a comparable beaded cap with fox teeth. In addition he had a belt adorned with over 250 polar fox canine teeth and an ivory pendant carved in the form of an animal. Beneath his left shoulder was a sculpture of a mammoth made from ivory, and nearby there was found a polished human femur filled with red ochre. To his right was an ivory lance 2.4 metres in length and fashioned out of a mammoth tusk. As the lance weighs several kilograms, White thinks it was unlikely to have been used as a functional weapon and it may thus be a purely symbolic object. The third burial is that of a girl age between seven and nine years old. Her beads are of the same size and design as those of the boy but even greater in number, totalling 5,274. Like the others she wore a beaded cap but the complete lack of fox teeth may possibly indicate that such items were associated only with males, although this cannot be said with any certainty. Her grave goods consisted of some small-scale ivory lances, two batons made from antler, one of which was embellished with drilled rows of dots, and three discs of ivory.

Such richly ornamented clothing is rare even today and its occurrence way back in the Stone Age shatters any belief that the people of this time were too primitive to fashion such garments and personal ornaments. Yet the presence of these thousands of beads can also tell us something about the nature of the society that made them. Russian archaeologists who have conducted experiments in which they reproduced examples of the ivory beads from Sungir concluded that each one would have taken about 45 minutes to make, and White believes this to be a conservative estimate. Even using these cautious figures for the manufacturing time, we can see that the making of the beads for the man must have taken over 2,000 working hours, and those of the children more than 3,500 hours each! Bearing in mind that the various ivory articles that were among the grave goods of the children would also have taken time to make, a major investment in labour is clearly indicated. It seems highly probable that such lavish burials were not given to all members of the society at Sungir and that these individuals were of high status. Clearly not everyone could have thousands of beads made for them, and the fact that the two most complex burials were those of children makes it clear, as White points out, that not only did a hierarchical society exist at Sungir, but that 'social position was ascribed rather than achieved'.

Further proof that the early Upper Palaeolithic was not inferior to the later

Upper Palaeolithic in its artistic achievements is embodied in a figurine of Aurignacian age, the first parts of which were found on 23 September 1988 at Galgenberg, near the town of Krems in Lower Austria, at a site excavated by Christine Neugebauer-Maresch. The figurine (see *Figure 32*) was found broken into a number of scattered fragments which were put back together by its excavators to reveal the representation of a woman who appears to be dancing. The archaeologists who discovered the object called it 'Fanny, the dancing Venus of the Galgenberg', and it is sometimes simply referred to as the Venus of the Galgenberg or the Galgenberg figurine. This curious name requires some explanation. The figurine was named after the Austrian ballerina of the nineteenth century, Fanny Elssler, and the description of the object as Venus is a standard epithet given to Palaeolithic female figurines, particularly those dating from the Gravettian period. Not only is the 31,000-year-old Galgenberg figurine much earlier than its Gravettian counterparts, it is also the earliest known female figurine in the whole of Europe and, according to conventional archaeological thinking, the world (although, as will become clear later on in this book, there is a far earlier example from the Near East). The Galgenberg figurine is made from a green stone and is seven millimetres thick and just over seven centimetres long. It weighs a mere 10.8 grams. Bednarik's comments on the skill required to make the figurine are instructive:

Technologically the production of the sculpture is significantly more advanced than that of any Gravettian (and thus more recent) figurine. The soft stone [out of which various Gravettian figurines were made] can be readily fashioned with flint tools but in this case the object is rather brittle and delicate . . . The several salient parts (left arm and breast, head) could all easily fracture at their base, and to crave or bore the two openings (between torso and right arm, and between legs) involved a very delicate process.

The stone's physical properties would not permit the fashioning of a free-standing limb, especially an arm. To overcome this limitation the artist utilised two different conventions still being used by contemporary sculptors: the right arm and the legs are structurally supported (and thus braced) at both ends, while the left arm is shortened to half the anatomical length by being depicted in a folded-back position. This alone shows that the artist was well versed in the techniques of producing human figures with 'free' limbs, an art that was apparently not mastered by the Gravettian artists. Such advanced skills demand an accumulated store of artisan's know-how and cannot be explained as anything but the product of a lengthy tradition in which people had experimented for thousands, and probably tens of thousands, of years (perhaps with perishable media [such as wood] ?).

The extraordinary skill of the Galgenberg artist is also shown by his or her ability to maintain a definite and vivid visualisation of the intended

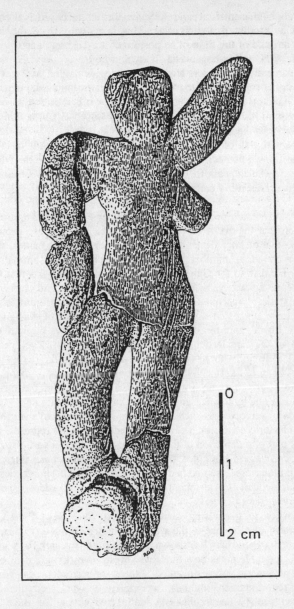

Figure 32

form throughout manufacture, despite the various technological challenges involved in producing the figurine. This is evident from the internally coherent attitude of the figure: the posture of all the body parts is correctly balanced with the whole. The body's weight is depicted as being supported mostly on the left leg; the right leg is angled and resting on a slightly higher support than the left. This facilitates the casual placement of the right hand on the upper thigh. The upper torso is therefore turned to the left, a position also demanded by the steeply raised left arm. This attitude brings the left breast almost into profile, showing it to be large and consistent with that of a young woman. The second breast is in low relief, due to the stone's flatness. Facial detail is lacking, and while the wide upper part of the head appears to be so shaped intentionally, we cannot know whether it represents a coiffure or is merely incidental.

The skill and vibrant power of this early Upper Palaeolithic figurine refute any notion that the art of the Aurignacian was in any sense primitive, or inferior to that of later Upper Palaeolithic art or, for that matter, the art of any other epoch. A number of Aurignacian carvings of animals have a similar dynamism to the Galgenberg figurine, which suggests that this kind of representation was very much part of their tradition and clearly distinct from the static representations of the Gravettian Venus figurines.

It has been suggested that another important aspect of Aurignacian art was their liberal and frequent use of sexual imagery, particularly their depiction of the female genitalia. This theory was first developed by the same l'Abbé Breuil who was later to take an interest in the sexual nature of the Ain Sakhri figurine. The idea soon caught on among French prehistorians and became something of a dogma, and various shapes engraved in stone by the Aurignacians that vaguely looked like a vulva were automatically perceived as such by scholars eager to discover further proof of the prehistoric obsession with sexual matters. Whilst representations of vulvae as part of a depiction of the whole or a substantial part of the female body are convincing examples, most images that have been interpreted as female genitalia are disembodied and usually of such simple design that they could equally plausibly have been made to depict something completely different.

In a witty and amusing article entitled 'No Sex, Please, We're Aurignacians' (a reference to the well-known British farce *No Sex, Please, We're British*), Paul Bahn has shown how absurd this search for such sexual images became. He notes how those who were investigating possible cases of the depiction of vulvae invented a series of ridiculous categories and types of representation such as 'incomplete vulvae', 'circular vulvae', 'trousers vulvae' (whatever that is!), 'unfinished vulvae', in order to force the diversity of images which were clearly not vulvae at all into the picture of Aurignacian sex mania. Perhaps the most absurd example of all is a description of a simple straight line as a representation of the vaginal opening. One can only think that the sexual obsession is not Aurignacian in origin but rather derives from the prehistorians themselves. This is even more apparent when considering some of the views expressed on the more general role of the Venus figurines in later Upper Palaeolithic cultures. Bahn quotes with not a little disdain the view of B. Kurtén that such images unequivocally had a direct sexual purpose:

> Female figures often appear in sexually inviting attitudes, which may be quite the same as those seen in the most brazen pornographic magazines. There are also anatomically detailed pictures of the vulva, showing the female sex organ sometimes frontally, sometimes inverted and from the back, open to penetration.

Kurtén was not the first to suggest that the Palaeolithic depiction of females was primarily for sexual gratification. In the 1940s Karel Absolon, the excavator of Dolni Vestonice, believed that the art found at this Gravettian site reflected the pornographic imagination of its makers. Among the articles discovered at the site which he sees as embodying such interests is a carved ivory baton displaying what are usually described as breasts but could also be interpreted as testicles. Recently the archaeologist Tim Taylor has suggested that the object may have had a more direct masturbatory function, namely as a dildo. He suggests that the use of such objects could either have been purely for pleasure or have had a ceremonial function in the ritual deflowering of virgins. He cites the existence of other potential dildoes among Ice Age artefacts. The function of an enigmatic group of objects known as *batons de commandement* has perplexed prehistorians; some believe them to be magical staffs whilst others think they may have had a more practical function as spear-straighteners. As many of these objects are undeniably phallic, Taylor suggests that they too could have been used as dildoes.

He also thinks that a Venus figurine from the site of Kostenki I in Russia is sexually explicit. The breasts of the depicted woman are exposed and her hands are seemingly strapped behind her back, according to Taylor, and thus he interprets this as a possible indication of sexual bondage. In making this suggestion he states that he has not seen a picture of the rear view of the figurine and so does not 'know where the straps go around at the back, but the arms are being held firmly down behind'. This idea of prehistoric bondage and sado-masochism may be original and controversial but it is certainly wrong, at least in the case of this particular figurine. Even when the figurine is viewed from the side it can be clearly seen that the hands are not behind the back at all but hanging free at the woman's side. From the front view the lower part of the arms, including the hands, is obscured by the large breasts and so their placement could be misconstrued, as it has been by Taylor. The 'bondage straps' of this prehistoric 'sex slave' are better seen as armbands and bracelets.

Although some of the ideas of Breuil, Absolon, Kurtén and Taylor on the pornographic and overtly sexual nature of various art objects and engravings from the Palaeolithic period are, to say the least, rather dubious it would nevertheless be wrong to go to the opposite extreme and portray the prehistoric peoples of this time as puritanical. In fact there are a few straightforward and unambiguous examples of sexual themes such as a very basic 20,000-year-old drawing of a penis inserted into a vulva incised in stone at the site of Isturitz in the French Pyrenees. This very simple kind of work requires no more skill than that displayed by 'artists' in public lavatories. One should not jump to the conclusion that *all* prehistoric representations of sexual themes necessarily had ritual or magical connotations. Many of them could well have been simply 'art for art's sake' or even the Stone Age equivalent of modern graffiti for the idle doodler. Yet there are other more complex artistic renderings that contain sexual themes but obviously have symbolic overtones.

Whilst Bahn has certainly drawn attention to the extremes to which the 'vulvae hunters' have gone in their interpretations of prehistoric symbolism, there nevertheless remain a considerable number of genuine representations of vulvae in Upper Palaeolithic art. But this does not mean that they are sexual in orientation in all cases, and it seems very likely that in many instances the vulva is not depicted for its role in sexual pleasure but rather because of its associations with pregnancy, childbirth and menstruation – in short, it is not simply a sex object. In the Cougnac cave in Quercy, France (mentioned in Chapter 13), there is a natural cavity which suggests the shape of the .vulva, and this similarity of form was apparent to the prehistoric people who frequented the cave and stained it with red ochre to symbolise the menstrual flow rather than displaying any erotic connotations. A number of clefts and small cavities at other cave sites have also been marked with ochre for the same reason. Art that most probably dates from the late Upper Palaeolithic period was found in 1980 in the Ignateva Cave (Yamazy-Tash) in the southern Urals, Russia. Among the images is a female figure with 28 red dots between her legs, a depiction which seems clearly to refer to the menstrual cycle. A group of carvings from Kostenki II are incised representations of vulvae and also seem to show the flow of menstrual blood. It is clear even from these few cases that artistic renderings of the female genitalia are by no means always an indication of sexual desire but rather refer to other aspects of life. The same is true of the Venus figurines, most of which are hardly erotic at all, unless the mere fact that they represent women (and not always in a state of undress) makes them sex objects.

Most of the so-called Venus figurines belong to the Gravettian period and are sometimes described as if they were all part of a cultural complex which involved a shared ideology that spanned both the area in which they occur (from France in the west to Siberia in the east) and the many thousands of years in which the various figurines were made. Some researchers have seen the widespread nature of the Venus figurines as evidence of a goddess religion that was practised during this period across Eurasia, arguing that the virtual absence of male figurines from these times shows the essential role of the feminine in the spiritual life of the Gravettian cultures. To counter this, Paul Bahn has said that the widespread use of feminine imagery in the form of the Virgin Mary in Christian iconography obviously does not mean that Christianity is a goddess-worshipping religion. Thus even if the Gravettian figurines are mainly female, this does not automatically mean the religious life of this period of prehistory was based on goddess worship, nor that such worship was necessarily officiated by priestesses rather than priests. Pertinent as this view may be, there is, nevertheless, in Christianity a widespread iconography of the male figure of Christ, and no significant depiction of males appears in the Gravettian traditions. If the figurines are of such religious and ritual significance, then it seems likely to have involved the worship of a goddess rather than a male deity. The discoveries made by Marija Gimbutas concerning the religious life of Neolithic Europe support this view. Although there are a number of statuettes from Old Europe that may depict male deities (see, for instance, the statuette known as 'The Thinker' on the front cover of this book), the vast majority are clearly female. If, as Gimbutas believes, the Old European religion was dominated by the goddess in her various manifestations, then a similar situation may have also prevailed in the Gravettian period. Compared with the rich archaeological remains from Neolithic Europe, the Upper Palaeolithic evidence is rather sparse, and apart from the female figurines, there are few artefacts from the period that can shed light on the religious beliefs and practices of this remote time. As a consequence there is still much controversy about the meaning of the Palaeolithic figurines.

Some researchers say that it is simply impossible to decide whether the figurines represent a goddess or goddesses at all, and that they may rather depict real women. Bednarik is very sceptical about the usefulness of lumping all female figurines of the period together, noting that they are extremely diverse in numerous ways. For example, some are naked; others partly or fully clothed. Some are in a pregnant condition; others are not. Some have enormous, pendulous breasts (see *Plates XX* and *XXI*), whilst many others have no breasts at all. Some are fat to the point of obesity (perhaps suggesting fecundity or plenty), whilst others are very slender (see *Plate XIX*). Beyond the fact that they all depict females and most come from the same period of the Upper Palaeolithic, they appear to have little in common.

Despite the considerable problems involved in interpreting the meaning of the Venus figurines, it seems, nevertheless, that certain points can be made with a degree of confidence. Firstly, the fact that the majority of figurines known from both the Gravettian period and also the Upper

Palaeolithic period as a whole are female must have some symbolic significance. Secondly, it is clearly erroneous to see these figurines as having a widespread role as prehistoric pornography or even as erotic art. Some of them may well have been seen by their (male or female) makers as emblems of sexuality, but there are clearly many (such as those in which women are depicted in full clothing and without any emphasis let alone exaggeration of the erogenous zones) that are hardly likely to have had such a function. Thirdly, the fact that the figurines are found across a huge geographical area and a period of thousands and thousands of years means that it would be ridiculous to think that they all symbolised the same thing to their extremely diverse makers. It is quite apparent that the female body was a symbol used to express numerous concerns in Palaeolithic times.

A closer investigation of one of these figurines will illustrate the probable diversity of the symbolic meanings with which even a single work was invested by the Palaeolithic artist and fellow members of his or her community. Probably the most famous of all the female figurines of the Old Stone Age is the so-called Venus of Willendorf (see *Plate XXI*), a Gravettian work discovered at Willendorf in Austria in 1908. Although this limestone figurine is a mere 11 centimetres high, the carver has embellished it with a great deal of detail and it is rightly seen as a masterpiece. The most striking features of the head are the absence of the face and the presence of an elaborately carved coiffure. Both these features are not peculiar to the Willendorf figurine, for many – but by no means all – other female figurines lack facial features, whilst the depiction of a coiffure is by no means unusual in Upper Palaeolithic art. Other examples of coiffures include the bun and long tied braids, which show that the human preoccupation with using the hair to express cultural attitudes and styles was as much a concern in Palaeolithic times as it is today.

Alexander Marshack believes the coiffure of the Willendorf figurine may be one of the symbols of the mature and fertile woman. He has made a special study of the symbolism of the female during the Upper Palaeolithic and has used this particular figurine as his type image, partly on account of her geographical and chronological position. Geographically (and, says Marshack, stylistically) she lies between the western European 'Venuses', the southern 'Venuses' (of Italy) and the eastern 'Venuses' of Russia and Siberia. Chronologically she is preceded by the Aurignacian age Galgenberg figurine and the vulva images of the same period, and followed by female figurines which are, in some cases, of a more abstracted and schematic style. The vulva as a symbol, which as we have seen was a prevalent image in the Aurignacian period (although, as Bahn has demonstrated, not as all-pervasive as some have perceived it to be), is prominently portrayed on the Willendorf carving, which has, according to Marshack, 'the most carefully and exquisitely carved realistic vulva in the entire European Upper Palaeolithic'.

The ritualistic nature of the Willendorf figurine is suggested by the fact that it was originally covered with red ochre, which, as we have already seen, had a widespread symbolic function in numerous Palaeolithic contexts. The emphasis of certain feminine features such as the coiffure and vulva, and the corresponding lack of other features such as the face and the lower legs and feet are aspects of the figurine that for Marshack are not simply an expression of artistic style on the part of its maker but also likely to be the deliberate and conscious use of certain aspects of the female form with symbolic import. For him, this and other figurines as well as other works of prehistoric art are not simply art for art's sake but an integral expression of cultural values that he links with his theory that much Upper Palaeolithic symbolism was based on 'time-factored' concerns:

> If, as I suggest, these different symbol systems . . . were parts of complex conceptual systems that cohered and integrated divergent processes and periodicities – including the changing seasons, the flood and the thaw, the seasonal behaviour of animals, the periods and processes of human reproduction, the complex processes of human subsistence activity and movement, the periodic aggregation and rituals of human groups, the periodic use of the sanctuary caves . . . as well as the observable astronomical processes and periodicities noted in the sky – into a functional, if mythologized and metaphorical, cultural frame, we can begin to discuss the different systems that appear in the iconographic record, not as 'art', as animal images, as female images, as cave images, as personal decorations, as signs, etc., but as aspects of an encompassing cultural 'ethos' and an integrated fabric and frame of reference. The female image was probably, in this suggested context or framework, a symbol and a metaphor that encompassed a diverse range of meanings and possible uses; it was probably related to different aspects of the 'time-factored' cultural tapestry, involving the broadly biological and the human. The 'Venus' of Willendorf, then within *her* culture and period, rather than within ours, was clearly richly and elaborately clothed in inference and meaning. She wore the fabric of her culture. She was, in fact, a referential library and a multivalent, multipurpose symbol.

We can now see how any crude explanation of the Willendorf figurine as simply a fertility figure or an object of sexual desire is entirely inadequate. The representation of the female body during the Upper Palaeolithic period was not simply a means to express the sexual allure of a naked woman, nor was it merely a lucky charm superstitiously fashioned to aid fertility. Rather the female body was a symbol of cosmological significance that was able to express all aspects of Palaeolithic human concerns. Neither can the cultural expressions of clothing, coiffures and personal adornments such as bracelets be always taken simply at face value, for they can all be imbued with symbolic meanings. If it is indeed correct to see the female figurines as rich in symbolic and social meanings and also to perceive them as having had, at least in many cases, ritual functions, then a picture emerges of a Palaeolithic body of religious and practical knowledge that was expressed in the various parts and attributes of the female form – in representations of pregnancy, in the carving of the breasts and the vulva, in the form of the coiffure and in other aspects of this rich and complex means of expression. If the female body was one of the most widespread and elaborate images of the Old Stone Age, and a symbol for the various forces of nature and the various aspects of culture, would it really be so far from the mark to believe that the figurines actually embody aspects of the Palaeolithic worship of a goddess?

Chapter 15

The Song of the Stalactites

During routine sand-quarrying that took place at Schulen in northern Belgium during 1976, a number of stone artefacts and animal remains came to light, and some of them were salvaged by amateur archaeologists. A local schoolmaster named Mr Vangeel was among the enthusiasts, and he collected an engraved mammoth long bone that has been dated to the Middle Palaeolithic period, about 50,000 to 40,000 BP. The dimensions of the object (*Figure 33*) are as follows: maximum length 31.2 centimetres, maximum width 8.7 centimetres and maximum thickness 2.4 centimetres. One end has been fashioned into a point by the use of a sharp tool, and the marks caused by this process are clearly visible to the naked eye. The point of the bone and the area beneath it have been polished, and according to Dirk Huyge, an archaeologist who has made a special investigation of this artefact, this polish is easily distinguished from the natural sheen of bone. The most interesting signs of human work are a number of incised and roughly parallel lines extending from the area close to the point downwards towards the main part of the object. There are twelve grooved lines discernible on the bone, although it is possible that additional lines may have existed and have been worn away. The regularity of the lines indicates that they are due to deliberate human action rather than being bite marks caused by a carnivorous animal.

Huyge has proposed that this object had two uses, which were unrelated. Its initial use seems to have been related directly to the presence of the parallel grooves, whilst it was subsequently used as some kind of pointed object. This latter usage was the cause of the end polish and the disfigurement of some of the lines, perhaps, as said above, to the point of obliterating some additional lines. Huyge thought that the original use of the object could have been strictly utilitarian; for example, it could have been a thong stretcher. He rejected this possibility as an unlikely one because this kind of use would have caused wear and polish of the grooves. This had not occurred as traces of the action of the tool that had cut the grooves were still discernible, whereas if thongs had been repeatedly placed in the grooves they would have covered up the traces of the original cutting action. In fact the ribs (i.e. those parts that projected up to the original surface level of the bone and were between the grooves) had been made glossy probably as a result of having been struck repeatedly by another object. Huyge has found a more colourful explanation of the artefact, far removed from the prosaic thong stretcher:

> In my opinion a valid alternative explanation of the function of the bone may be that of a sound-generating device. Indeed, the well-formed groove pattern endows the object with some resemblance to a certain type of musical instrument, a rasp or scraper. The rasp is a primitive ... instrument with a corrugated surface that is scraped rhythmically by a non-sonorous object to produce sounds. Striking the parallel grooves of the Schulen bone with a rigid instrument, such as a wooden rod or a bone plectrum, indeed produces a ... sound. However, if the object was indeed used for producing sound in this way, that sound would have almost certainly differed from the one it can produce today. The alteration of the shape of the bone and its fossilisation may have profoundly changed its acoustic characteristics.
>
> Rasps or scrapers are well-known instruments, both from archaeological and ethnographic sources. Their distribution is almost world-wide. In its simplest form the rasp consists of a notched stone, bone, shell or gourd, which is scraped with a stick or other rigid object, the sound being increased in some instances by placing the instrument over a hole in the ground. It may be considered, together with the flute, the lithophone [for which see below] and the bullroarer, as one of the earliest musical instruments known to man.

Few instruments are so simple either to manufacture or to play and, as Huyge points out, the scraper is reinvented on a daily basis by schoolboys who run their rulers along fences to produce noise and rhythm. The possibility therefore exists that Neanderthal people may have made at least simple musical instruments, and if they did so then the rasp would be among the more likely candidates. As will become clearer when other pre-Upper Palaeolithic artefacts are described below, many engraved bones display a regularity in their markings which shows the ability of these early humans to represent regularity in visual terms. That they may also have expressed this capacity by means of sound is by no means far-fetched particularly when we consider the observations of Boris Frolov on the origins of counting and its relationship with the rhythmical action of tool-making from the Lower Palaeolithic onwards (see Chapter 6). Despite the possibility of a Middle Palaeolithic musical tradition, the evidence is, nevertheless, slim. Although a mammoth tooth artefact from the Mousterian site of Tata in Hungary has been interpreted as the remains of a bullroarer (a musical instrument consisting of a piece of wood attached to a cord and swung around in the air to produce sounds), this is not likely, particularly as it lacks a suitable hole or other means of attaching a cord. Hollow bones with holes in them have been put forward as Middle Palaeolithic 'whistles' or 'flutes'

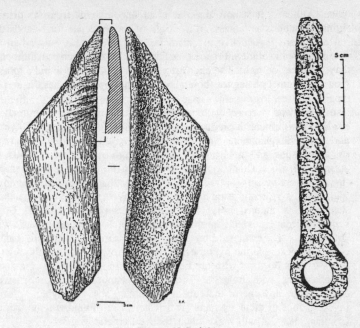

Figures 33 and 34

but dismissed as simply the result of animal activity, the holes being the result of penetration of the bone by teeth.

Philip Chase, one of the most vocal critics of the existence of Middle Palaeolithic symbolic capacities, made a first-hand investigation of the animal remains from the Combe Grenal rock shelter site in the Dordogne, France, and found that although some of the pierced bones did produce sounds, many of those that did could not possibly have been functional whistles. Many of the bones displaying holes showed tooth marks on the opposite side, thus indicating that the holes were simply puncture marks caused by animal activity. Vadim Stepanchuk, co-excavator of the Middle Palaeolithic Prolom II Cave site in Crimea discovered by Y.G. Kolosov, has found evidence for what he sees as symbolic activity by Neanderthals who used the cave. At the back of the cave he found a concentration of large bones, jaws, tusks and other remains belonging to a variety of animals including mammoth, cave bears, giant deer and bison. The bones were deposited in an arc shape about eight metres long and a metre wide. Unlike the rest of the animal remains in the cave which had been broken into small pieces and were scattered in disarray, this concentration of bones appeared to Stepanchuk to be the result of some intentional activity with no apparent practical or utilitarian function. He concluded that it most likely had a ritual purpose. Although it is now popular in archaeological circles to dismiss such interpretations as fantasy, in the light of the evidence for Neanderthal burial and more particularly the artificial structure recently found in the Bruniquel cave in southern France (see Chapter 10) there is a possibility that Stepanchuk may be right. The excavators of Prolom II also found a total of 111 pierced animal phalanges (bones), of which Stepanchuk has this to say:

> The origin and purpose of these holes is not quite clear. The study of phalanges with holes has already been going on for more than 150 years, and various explanations have been proposed: the obtaining of marrow; use as whistles; and the result of biting through by a carnivore whilst the animal was alive. Other hypotheses seem to be fantastic, for example, that they were vessels for poison. It is possible that some of the phalanges with holes were really used as whistles. R. Wetzel wrote that phalanges with roughly pierced holes from [the archaeological site of] Bock-steinschmiede H. which he had recognised as 'hunters' pipes' were shown by experiment to utter quite strong shrill sounds. One cannot completely exclude the hypothesis about marrow procuring, although in many ways it does not withstand criticism . . . In any case, the abundance of phalanges with holes at Prolom II cannot be comprehensively explained by any one of the causes mentioned above. Maybe . . . future investigations of these artefacts [which have been found at a number of] Crimean sites . . . will make clear their enigmatic origins.

Whether or not the Neanderthals made whistles, such practices may have originated not with the desire to play music as such but to imitate birds or other animals to act as a decoy in hunting.

Archaeologists have had less of a problem accepting the existence of musical instruments during the Upper Palaeolithic period, including a Magdalenian bone artefact from Pekárna Cave in Moravia which is most probably a rasp (see *Figure 34*). Other rasps and scrapers are known from later periods of prehistory – for example at Neolithic sites in the Near East and the Early Iron Age site of Janakowo in Poland. There are reports of some 30 cross-blown flutes or whistles with between three and seven finger holes, found from France to Russia, mainly dating to the period from 20,000 to 15,000 BP, and made of bird, bear and reindeer bones. Some such whistles from the Magdalenian period are decorated with chevrons and other designs which indicate that such objects cannot be confused with bones gnawed by carnivores. Without doubt the most spectacular collection of Upper Palaeolithic musical instruments that has been reported is that discovered in an Ice Age house built from mammoth bones at Mezin in the western Ukraine, dated 20,000 to 17,000 BP. It is a whole set of percussion instruments made from mammoth bones – skull bones, hip bones, shoulder blade, jaw bones – decorated with ochre, and accompanied by a 'castanet bracelet' and two rattles made of ivory. These finds have been described as evidence for a 'Stone Age Orchestra', an orchestra that played on mammoth bones in a mammoth bone house and no doubt were given mammoth flesh to eat after their performances. In what must be a unique instance of experimental archaeology, a group of archaeologists not only played these instruments but even recorded their efforts on a record!

There also appears to be a musical aspect to the decorated caves of the Upper Palaeolithic period. The Abbé Glory was the first researcher of the caves to pay particular attention to the possibility of playing the caves themselves – or at least parts of them. He reported that certain caves had natural percussive properties and that when folds in the walls were struck they gave off evocative sounds. These vast instruments could have been played by striking them with a flint, bone or wooden object as a plectrum is used to strike guitar strings. He gave such ready-made instruments the name lithophones. He surmised that prehistoric man at the site of the Roucadour Cave in France had made use of such properties in his ceremonial activities, stating that a:

> rite seems to have been accompanied by rhythmic sounds. As speleologists [specialists in the study of caves] know, some stalactites vibrate and emit sounds when struck ... prehistoric man was acquainted with this effect. At the cave of Nerja, near Malaga ... edges of folded draperies were chipped of old by alternate blows on both sides. We have experimented with some and obtained deep sounds like a tom-tom.

Having read about Glory's pioneering investigations and experiments into this peculiar Palaeolithic practice, Lya Dams decided to take the matter further and conduct her own detailed investigations at the Nerja Cave. This Spanish cave system consists of two main parts known as Nerja I and Nerja II. The lower level, Nerja I, is highly popular with visitors and an estimated 350,000 people visit it each year. Its acoustic qualities are attested to by the fact that a series of summer concerts are held there. The Nerja site was used as a burial site during the Neolithic and Bronze Age periods, but Upper Palaeolithic (Solutrean) burials with accompanying red ochre have been found at the back of Nerja I, in the area known as the Hall of the Cataclysm due to the violent geological actions that racked the cave in the remote past. Lya Dams and her husband spent a number of years documenting more than 500 Palaeolithic paintings and engravings in the cave, most of which were of abstract forms rather than figurative depictions of animals. Whilst undertaking this painstaking work they realised that some particular groups of designs in both parts of the cave system were unusual enough to be given special status and designated as 'sanctuaries'.

It is not just the presence of the artwork that characterises the site as a sanctuary; it is also due to a number of other features, such as the striking nature of the location itself, the restrictive space which precludes entry to the place by more than a few individuals, and the sheer effort involved in reaching the actual area to be painted, which may require scaffolding if, as is sometimes the case, the selected zone is high above the floor of the cave. The Dams believed that the inner sanctuary of the Nerja site was a recess in the Hall of the Cataclysm to which they gave the name 'The Organ'. 'The Organ' has the appearance of a triangular platform and is invisible from the cave floor. The right-hand wall of this sanctuary consists of a series of folds in the rock which are about three to four metres in height and separated by a space of a few centimetres in between. In Palaeolithic times access to 'The Organ' would have been achieved by scaling a number of large boulders. Lya Dams describes the workings of the instrument in this natural cathedral:

> Various sounds can be obtained when striking the folds' edges with a hard object, hence the name of the recess. We have experimented with blunt flints and wooden sticks; the clearest, most harp-like and resonant notes echoing to the rear of the cave have been obtained when handling the end-folds with the latter. Several edges have been intentionally broken of old at different heights; this may have been done to vary the sounds, because we assume that the cave-artists used these edges, which have a quite worn appearance, as a huge percussion or lithophone, accompanied perhaps by a flute, clapping of the hands or stamping of the feet.

The esoteric nature of this area is indicated by the fact that complex symbols were painted (mostly with red ochre) in both the folds and the grooves between them. Most of the figures are almost invisible at first sight and were clearly not put there for the purpose of public display. Only one figure – which Dams describes as a signpost or beckoning sign – is at all visible from a distance. She suggests that the symbols depicted in 'The Organ' may be a complex notational system relating to ritual activities and that they could have acted as a mnemonic code to remind the 'organist(s)' of details of ceremonies which are now long forgotten. 'The Organ' and the other known cave sites of lithophones in Spain, Portugal and France have, despite the grandeur of their scale, a certain resonance with the humble rasp known from Magdalenian times and may perhaps echo an even humbler origin in the Mousterian era. Other investigations are under way on the acoustic properties of Upper Palaeolithic caves with claims that decorated areas are also often the areas in which the male human voice has its optimum resonance. As Paul Bahn, one of the leading archaeological authorities on the cave art, has said: 'it is extremely likely that they [the Palaeolithic users of the caves] would have used any acoustic peculiarities to the full'.

The so-called Venus of Laussel has already been made mention of in relation to the astronomical and calendrical activities of the Old Stone Age, but this enigmatic bas-relief (see *Plate XVIII*) is also open to a number of other interpretations, including musical ones. One theory suggests that the object held in the right hand of the woman is actually a horn or some other wind instrument, but as neither the end nor the side depicts a hole for blowing it, this seems unlikely. A more recent idea has been proposed by Dirk Huyge, who suggests that, like the Neanderthal object discussed above, it is another instance of a rasp or musical scraper. In many tribal cultures the rasp is an instrument that is imbued with magical overtones and for this reason Huyge suggests that it may have had a ritual significance, perhaps as a magical fertility aid to pregnant women in Upper Palaeolithic times. Another kind of musical instrument which can be traced back to Stone Age times is the conch or marine shell. Triton shells have been found at a number of Neolithic and Chalcolithic sites, such as those belonging to the Lengyel culture of eastern Europe. The blowing of a conch has had a ceremonial function in numerous parts of the world. It still fulfils such a role in the rituals of many Melanesian societies and retains its importance even in contemporary Hinduism and Buddhism. Other kinds of instrument that date from the Neolithic period in Europe include rounded ceramic whistles with a single blow hole (an instrument highly reminiscent of the ocarina) and enclosed rattles made of clay.

Various kinds of drum were also in widespread use in Europe at this time. Among a number of miniature ritual clay objects found at the site of Ovcarovo in northern Bulgaria (dated around 4500 BC) were three cylindrical drums. Although the drums themselves were not decorated, many of the other objects such as ceremonial bowls, altar screens and figurines were embellished with meandering lines and spirals. It seems likely that the model drums were replicas of full-scale drums that would have been beaten to accompany the religious ceremonies of this prehistoric community. There is also other evidence that points to links between drumming and religious beliefs in early times. Numerous Early and Middle Neolithic clay drums, ranging from large-scale bass drums to much smaller kinds, have been found at archaeological sites belonging to a culture known as the Funnel-Necked Beaker culture (the name derives from their distinctive kind of pottery). Over 60 have been found in Germany alone, half of them in megalithic tombs. That such drums could actually be played was shown when modern replicas were made based on the design of Neolithic examples found in Bohemia. The skins used on these modern copies were made of uncured leather, which was probably the original kind of membrane used by the prehistoric drummers. That the membranes were stretched to make them taut is known by the common occurrence of pierced lugs on the Neolithic drums.

Figure 35 shows an hour-glass-shaped Neolithic drum from the Polish site of Mrowino; the lugs are clearly visible. A similar drum, this time from Rössen in Germany, is shown in *Figure 36*. But unlike the Polish example, this one is decorated with four protruding pairs of breasts, one on each side of the drum. This unusual design may have had religious rather than sexual connotations, as well as serving a practical function as a means to attach the drum skin and also to tighten it. Prehistoric music by its very nature does

Figures 35 and 36

not leave as many traces as the highly visible arts of painting and sculpture, and only features rarely as a subject in either of these two artistic forms. Most instruments would most likely have been made from perishable materials such as wood, and as a consequence are unfortunately lost forever. Yet a faint echo of the range of musical expressions that may have existed in the Stone Age can nevertheless reach us across the ages, from the 'organ' of the caves – essentially a rasp on a grand scale – to the bone scrapers of Upper Palaeolithic, and perhaps even Middle Palaeolithic times.

Chapter 16

The First Fossil Hunters

In Chapter 13 we have seen how art appears from the very beginning of the Upper Palaeolithic period. After looking at the origins of writing it became clear that there was a prehistoric heritage of the use of signs long before the 'sudden' appearance of Sumerian writing. This should make us suspicious of the apparently equally 'sudden' birth of art and symbolic behaviour. As will become apparent, the ideological wall that has been placed at around 40,000 years is, like the wall that has been placed at 5,000 years ago to represent the beginning of history, civilisation and writing, also liable to fall when certain anomalous cases of earlier symbolic and artistic activities are brought to light. Here the stakes are, in some respects, even higher, for whilst the edifice built around the birth of 'civilisation' was designed to keep barbarians and primitives at bay, the 40,000 BP line for the birth of art is a means by which the earlier hominids – Neanderthals (see *Plate XXV*), archaic *Homo sapiens*, and *Homo erectus* – are excluded from involvement with the creative 'explosion' which some would have us believe developed almost magically, and with no precedent (nor with any hard evidence to show that some biological evolution occurred at this time).

I shall show that there are not simply one or two possible cases for symbolic activity by hominids before 40,000 years ago, but a much greater number, too many, in fact, to be swept under the carpet. I shall present them as exhibits for the defence of the idea that the art of behaviourally modern humans grew out of a long tradition of symbolic activity that can be traced back to an extremely remote period. Each of these exhibits (or groups of exhibits where they are best discussed as a group rather than individually) will be accompanied by full details concerning the objects or practices in question, their dates, where, when and by whom they were found, and an account of the debate – for and against – surrounding their status as evidence of symbolic behaviour. Not all of these exhibits can be defended adequately and the reasons why they must be rejected, or judgement suspended pending further investigation, will also be given.

In common parlance the word Neanderthal is used as a term of abuse for a person who is crude, ignorant and indifferent to the higher things in life, one unable or unwilling to develop any sense of aesthetic awareness. But how does the life of the Neanderthals of popular imagination fit with the real Neanderthal? Were they simply beasts driven by their instincts and using body and mind to pursue only practical and utilitarian goals, or did they have artistic and intellectual interests like we do? Many archaeologists share in the popular view of Neanderthals and of course have partly inspired and fuelled it. Others paint a more flattering picture of the extinct members of our lineage and accredit them with more mental and linguistic capacities. There are clues that show that aesthetic awareness and many other modern human traits of a non-utilitarian nature can be traced back to the Middle Palaeolithic period and beyond into the Lower Palaeolithic, thus undermining the theory of the cultural 'big bang' that is supposed to have occurred at the beginning of the Upper Palaeolithic period around 40,000 years ago. The archaeological record of the Lower and Middle Palaeolithic periods is, as I have had reason to mention earlier, dominated by stone tools, and it is the study of such artefacts that provides one way to gain a glimpse into the mental and other capacities of hominids. As the kinds of general information that can be obtained by the study of such tools (cognitive abilities, language capacities, etc.) have been dealt with earlier in the book, I shall not repeat what I have said there but just touch upon what they can reveal to us about aesthetic awareness in these remote times.

Fossil hunting is by no means an innovation of the modern world; the collection of such objects by professional palaeontologists and enthusiastic amateurs is foreshadowed by a human quest for the curious that goes back to remote prehistory. Siberian peoples who came across the frozen remains of mammoths made ivory carvings from their finds, and the Chinese search for 'dragon's bones' and other reputedly medicinal fossils is well known. In fact, the earliest hint that China had once been the home of early man was due to the discovery of a fossil tooth by the German palaeontologist Max Schlosser in a Chinese apothecary shop at the turn of the century! Fossil man himself shared this interest, as the discovery of a number of items makes clear. Among the curios collected by Neanderthals at the site of Arcy-sur-Cure in France was a fossil gastropod shell. In Afghanistan, the Mousterian level at the Darra-i-kur site contained a fossil shark tooth which according to its discoverer Louis Dupree shows signs of having been modified by Middle Palaeolithic man.

Alexander Marshack has reported that Derek Roe, the Director of the Donald Baden-Powell Quaternary Research Centre at the University of Oxford, informed him that fossil shells found in Bedford, England, dating from the Lower Palaeolithic period had been cited as an extremely early example of beads. Marshack went to Oxford to examine the shells under the microscope and found no evidence to suggest that the natural holes in them had been artificially widened. He did, however express the opinion that the remains of some organic material in the holes might possibly indicate that if not intentionally altered, the shells may nevertheless have been strung on a cord or thong and thus have had some symbolic or decorative role. From the Lower Palaeolithic period at least four Acheulian stone artefacts displaying fossils are known from Europe. One of them was brought to the attention of

Kenneth Oakley by K.D. McRae, who had noticed this unusual artefact among the hand-axe collection in a Cambridge museum. This singular hand-axe, found not far from the Brandon flint site in Norfolk, England, has a fossil bivalve mollusc shell in the centre (see *Plate XXVI*). It might be thought that, considering the enormous number of hand-axes that have been discovered, the presence of a fossil in this particular one might be purely fortuitous and that Acheulian man may not even have shown it any attention. Oakley thought otherwise, and noted that:

> The most significant fact which emerges from examination of the hand-axe is that the fossil shell distinguishing it is on a weathered portion of the block of flint out of which the implement was fashioned. We must infer that the Acheulian tool-maker not only saw the fossil when he selected the block of flint, but that as he worked this into a bifacial hand-axe he deliberately avoided flaking the area containing the fossil and left it occupying a central position ... I suggest that it was the fine fan-shaped markings of the Cretaceous fossil shell which appealed to the mind of the maker of the ... hand-axe, about a quarter of a million years ago.

If the Acheulians had an eye for the unusual, they also had an eye for perfection. Numerous archaeologists have made the observation that some hand-axes from the period are often fashioned to a level that seems to go beyond the fulfilment of utilitarian requirements; a number of them are works of beauty in their own right even without fossil decoration. From an even earlier period the unusual colour of tools made from cobbles of green lava at Olduvai Gorge has been seen as a possible indication of a primeval appreciation of colour. At the Olduvai site known as FLK North (Upper Bed I), Mary Leakey came across a peculiar cobblestone that had been grooved and pecked and for which she could not see any utilitarian function. No comparable artefacts from the whole of Olduvai Gorge have been found. On the upper surface of the stone there is an artificial groove which she notes was demonstrated by experiment to be deep enough for a thong or string to be attached. Considering her pre-eminence in Lower Palaeolithic archaeology, her comments upon it are of considerable interest:

> This stone has unquestionably been artificially shaped, but it seems unlikely that it could have served as a tool or for any practical purpose. It is conceivable that a parallel exists in the quartzite cobble (see *Plate XXVIII*) found at Makapansgat [a 2–3-million-year-old *Australopithecus* site in southern Africa] in which natural weathering has simulated the carving of two sets of hominid – or more strictly primate – features on parts of the surface. The resemblance to primate faces is immediately obvious in this specimen, although it is entirely natural, whereas in the case of the Olduvai stone a great deal of imagination is required in order

to see any pattern or significance in the form. With oblique lighting, however, there is a suggestion of an elongate, baboon-like muzzle with faint indications of a mouth and nostrils. By what is probably no more than a coincidence, the pecked groove on the Olduvai stone is reproduced on the Makapansgat specimen by a similar but natural groove and in both specimens the positions of the grooves correspond to what would be the base of the hairline if an anthropomorphic interpretation is considered. This is open to question, but nevertheless the occurrence of such stones at hominid sites in such remote periods is of considerable interest.

Another unusual discovery that may indicate a capacity for symbolic thought in a very remote period of prehistory is the Daraki-Chattan cave site in the Chambal Valley in India. On 8 September 1993 an archaeologist named Ramesh Kumar Pancholi was travelling in the region with some friends and discovered a number of cupules, or cup-shaped depressions, on a rock surface near a local temple. Believing them to be very old, he returned to the area on 2 October with his son in order to see if he could find other examples of these unusual phenomena and, as a result, discovered numerous cupules in a sandstone cave to which he gave the name Daraki-Chattan, meaning 'fractured rock'. The cave site was more thoroughly investigated by Giriraj Kumar and his son during December 1995. Daraki-Chattan is 8.4 metres deep and in places very narrow, particularly towards the back, where the last metre stretch is less than 40 centimetres wide. Kumar and his son were able to record the size and location of 498 cupules on both of the cave walls, and some of the cupules were discovered at a height of 3.5 metres from the floor, which indicates the effort that must have been involved in their execution. Despite the technical difficulties of using a camera in such a dark and confined space, the investigators were able to augment their diagrams and measurements with a photographic record.

Although the cupules are clearly made by humans (similar clusters of numerous cupules have been found elsewhere, for example in northern Australia and in Europe), assigning a date to them is very difficult. A number of stone tools of Acheulian and Middle Palaeolithic types were found on the cave floor, but may have been washed down by rainwater from the hill above, where comparable tools have been found. Such tools are far more common in this region than those belonging to the Mesolithic and Upper Palaeolithic periods. Kumar also found, about 1.5 metres outside the cave mouth, a series of five stones arranged in a line at a right angle to the mouth. These stones, and a further one close to the wall, may, according to Kumar, be the remains of the base of a burial structure or a wind break. He believes that, despite the lack of definitive evidence, the cupules in the cave and the remains of an artificial stone structure are the work of Acheulian people who once used the cave. He also suggests that the narrow passage in which the cupules were made would hardly have been the best place to display such work were it to be admired by all, and that the cupules may have had some ritualistic significance. There are a number of other sites in India for which claims of cupules and curved lines made in Acheulian times have been proposed, thus providing further hints of such activities on the subcontinent.

Quartz crystals also seem to have attracted the aesthetic attention of early man in a number of parts of the world. From the Lower Palaeolithic site at Singi Talav in India, six crystal prisms had been brought to the site by Acheulians; analysis indicates that they did not come from the same crystal flower and so were perhaps collected on separate occasions. Their small size (the largest is 2.5 centimetres long) makes it extremely unlikely that they could have been used as tools, and Robert Bednarik has suggested that they may have appealed to the Acheulians because of their visual qualities. About 20 pieces of quartz crystal had been collected by *Homo erectus* at the Zhoukoudian site near Beijing, and similar finds are reported from Acheulian sites in Israel and Austria. Interest in quartz crystals is evident in the Upper Palaeolithic period among the Magdalenians, who also collected fossils on occasion. There are reports of finds of quartz crystals at burial sites in California as early as 8000 BP, and many peoples across the globe attribute magical powers to them. In both North and South America they play a very important role in shamanism and are seen as a source of great spiritual and healing power. Similarly they are used in the magical and medicinal practices of various peoples in New Guinea and the Malay Peninsula. The Aranda people of Central Australia lacerated the body of the initiate with quartz crystals, whilst among another Australian people, the Wiradjeri, they were drunk with water to cause the consumer to see visions. We cannot, of course, know what use even the Magdalenians (not to mention the much earlier Acheulians) made of quartz crystals; whether they just perceived them as something nice to look at, believed them to be charged with magical properties or even had some practical purpose for them, we can only guess.

Although the presence of quartz crystals in the sites associated with early man may represent merely an aesthetic appreciation rather than indicating any developed spiritual aspirations, much firmer evidence for symbolic activity among the Neanderthals exists in the form of burials. Of all religious or spiritual activities, those concerned with burial rites are the most likely to leave a permanent trace for archaeologists to discover. There are no known burials from the Lower Palaeolithic period, although in the Atapuerca cavern system in Spain the remains of about 40 human individuals were discovered at the bottom of a deep shaft. The fact that there is no sign that the bones were the remains of a feast by carnivorous animals makes it reasonable to suggest that even if this may not be evidence of burial practices it may indicate the deliberate disposal of the dead. H. Ullrich has proposed that *Homo erectus* may have performed deliberate manipulations on the corpses of their dead, perhaps by disarticulation of the body or the skeleton, and suggests that the cut marks found on human bones at Zhoukoudian and other sites indicate such practices. Others have seen such marks as indications of the practice of cannibalism, perhaps with ritual overtones. Yuri Smirnov of the Institute of Archaeology in Moscow believes there is some evidence for a cult of skulls and mandibles (jaw bones) during the Lower Palaeolithic period because natural factors do not adequately explain why such a high percentage of skulls and mandibles (compared with other skeletal remains) have been preserved in the human fossil record. He also notes that throughout the Palaeolithic the preferential preservation of these parts of the skeleton is particularly striking in the case of children, which may indicate that such a cult, if it existed, paid especial attention to the young.

Until recently there were no generally accepted cases of intentional burial before the period 80,000 to 70,000 BP, but the discovery at the Qafzeh and Skhul Caves in Israel of burials dating from the period 120,000 to 90,000 BP indicates a greater antiquity for this practice than had previously been imagined. These burials contained the remains of humans very similar to modern *Homo sapiens sapiens*, which was something of a surprise bearing in mind the early dates. From about 70,000 BP, the number of known burials indicated that both the Neanderthals in Europe and western Asia (the Neanderthals of the latter region are sometimes called Neanderthaloid on account of their differences from their European cousins) and *Homo sapiens* (of a specific type described as 'Proto-Cro-Magnon') in the Near East were, at least on occasion, burying their dead. The number of known burials during the Middle Palaeolithic is extremely modest; Smirnov accepts about 60 cases which, for the period from which they come, is an average of two burials per thousand years. If these cases are accepted to be sufficient proof of the practice of burial – which many, but certainly not all, archaeologists believe – it is then clear that the burials that have been discovered are most likely to be only the tip of the iceberg.

One of the most dramatic Neanderthal burials to have been found to date was discovered in 1938 by a Russian team led by the archaeologist A.P. Okladnikov in a cave at Teshik-Tash in south-eastern Uzbekistan. It was on 4 July 1938, at the start of the excavations at the site, that the burial was discovered at the bottom of the uppermost cultural layer (Layer I). It turned out to be the last resting place of a Neanderthal boy aged about 12 and although the skull had been smashed into over 150 pieces by the weight of deposits above it, the team were able to reconstruct almost all of the cranium and facial bones of the skeleton. There is also some evidence that the body may have been deliberately 'defleshed' as part of the preparations for burial. The only bone tool at the site, a point made from an ulna of a Siberian mountain goat which may have been an awl or boring tool, was found close to the boy's body and has been interpreted as an offering to the dead. Five, or perhaps six, pairs of horns from the same kind of animal were arranged in a vertical position (or inclined towards the skull) with their pointed ends down in a roughly circular pattern around the skull and adjacent parts of the skeleton. Numerous other goat horns were found at the site – and make up about 84 per cent of all animal bones found there – but these were found lying horizontally and scattered in a random fashion. It seemed clear to the excavators that the goat horns found with the burial had been placed there deliberately and for a reason. It was suggested that the horns had some ritualistic significance and that a hearth found close by may also have been related to some funerary practices, perhaps along the lines of a feast, although this is very speculative.

In his analysis of the possible significance of the burial, academician Okladnikov noted that of the Neanderthal burials discovered before Teshik-Tash many were those of children and juveniles. Unlike others, he did not see this as evidence of the high rate of early deaths in Mousterian times, but rather as showing a particular bias by Neanderthals towards burying the bodies of children rather than adults, a trait which he found to be also true among some later cultures from the Upper Palaeolithic and Neolithic periods as well as among the modern Buryats (a culture related to that of the Mongolian people) of southern Siberia. He speculated that the purpose of the burial and the rituals that may have surrounded it was to ensure the rebirth of dead children. He also saw the burial as indicating a 'cult' of animals:

> The Mousterian 'cult of the dead' so definitely expressed in the archaeological finds in the Cave of Teshik-Tash; [indicates] a tendency to preserve the body of a dead fellow member of the human horde . . . Consequently among the Neanderthals the concept of life and death as being virtually distinct states must have already existed; otherwise it would be impossible to explain the Mousterian burial ritual, which in its essential features was identical for man and for those animals which served as the principal object of ceremonial activity in the Cave of Teshik-Tash (Mountain Goats) . . . For the deep and complex interweaving of the ancient Mousterian 'cult' of the dead with the 'cult' of animals is especially clearly attested by the finds in the Cave of Teshik-Tash, where the child's bones proved to be surrounded by Goat horns.

The existence of a cult of animals in which certain species (either highly prized game or dangerous carnivores in competition with man) play a key role in the ceremonial life of a society and are treated with special attention or even reverence (e.g. the collecting of their skulls) is hard to demonstrate for such a remote period. Okladnikov's ideas on the existence of such a cult are very intriguing but also highly speculative, and as such have been treated with considerable scepticism, although it is still generally agreed that the body at Teshik-Tash was buried intentionally by Neanderthals. That the mountain goat horns were placed in some deliberate pattern around the corpse has been questioned by Robert Gargett, who, in a 1989 article with the unfortunate title 'Grave Shortcomings: The Evidence for Neanderthal

Burial', has put forward arguments to show that *all* cases of reported Neanderthal burials can be explained by natural rather than cultural activities. Gargett claims that the 'circle' of horns was largely in the imagination of the excavator and that the whole case for Teshik-Tash being a site of deliberate burial is all based on misinterpretation and fantasy. Okladnikov's reputation as one of the world's most distinguished Palaeolithic archaeologists coupled with the fact that there are other reputable and convincing cases for Neanderthal burial has meant that Gargett's hatchet job has failed to cut deeply into either Okladnikov's reputation or the general consensus that there is more than sufficient evidence of burials from this time, despite certain shortcomings in the evidence from some sites.

The cave site of Shanidar in a remote part of northern Iraq has also yielded remains of nine Neanderthals, some in the context of burials and others as the result of accidental death due to an ancient rock fall. There are two particularly interesting aspects of the evidence from this site that may shed light on little-known aspects of Neanderthal existence. The first is the discovery of the mortal remains of a Neanderthal man aged about 40 who had died as the result of a rock fall about 46,000 years ago. This individual was given the official title of the Shanidar I Neanderthal although the expedition leader Ralph Solecki and his co-workers nicknamed him 'Nandy'. The skull of Nandy was the first part of his remains to be excavated, and it was the job of George Maranjian, the physical anthropologist of the team, to clean and preserve it as best he could in the rather basic conditions of the field laboratory. In his account of the Shanidar excavations Ralph Solecki tells of how an unwanted intruder nearly ruined Maranjian's work:

It was while the skull was exposed in the Shanidar laboratory, sitting in a tub of sand on the table in front of the window, that a small crisis developed. It was not unusual for chickens which the cook had destined for our supper to somehow find their way into the laboratory from the courtyard outside. In one brief unguarded moment one of these chickens had slipped in and made use of the sandbox in which the skull remains rested. We were shocked to see the chicken happily clucking to itself on the window sill, and some of the skull pieces, which Maranjian had been painstakingly cementing together, in disorder. Fortunately, Maranjian found that no damage was done, and the chicken was promptly secured, amid squawks, and given to the cook's care. We could not risk any repetitions of events like this. After that, unless the skull was covered, someone was always present in the room with it.

When the rest of the skeleton was removed from the ground it was transported with an armed Iraqi police guard by lorry and train to a laboratory in Baghdad for detailed analysis by Dr T. Dale Stewart. Stewart's study revealed that the right side of Nandy's body was withered – his right shoulder blade, his collar bone and upper right arm bone were underdeveloped, a condition which had probably been pronounced from birth. The indications were that during his lifetime the right arm had been amputated just above the elbow. He also suffered from a not uncommon problem among Neanderthals – arthritis. His teeth were worn down as a result of abnormal use, perhaps from the excessive chewing of animal hides to soften them, or as a result of using the teeth as a means of manipulation in lieu of his useless right arm. As if this catalogue of disabilities and ailments were not suffering enough, it was also found that he was blind in the left eye and had suffered and survived wounds to the face and skull. Clearly this individual must have been something of a practical burden to a group of mobile Neanderthal hunters, yet they had evidently looked after him as a part of their community since birth, as an individual in such a physical state could hardly have survived on his own. This shows that rather than being little better than a pack of wild animals, the Neanderthals clearly did not base their social beliefs around a 'survival of the fittest' kind of ethos but showed care and consideration to those who suffered physical disability. This level of social responsibility and conscience is all the more remarkable when one considers that there are many instances in historical and more modern times in which weaker individuals have, through necessity or cultural beliefs, been abandoned or neglected. Those who have read Jean Auel's popular saga *Earth's Children* will recognise in this account of Nandy, the crippled man of Shanidar, the source of her character Creb, the Mog-ur, or magician of the Neanderthal clan in *The Clan of the Cave Bear*, the first novel in the series.

A second discovery from the site that has caused much controversy and speculation is the so-called flower burial of the Shanidar IV adult male skeleton found some 15 metres from the cave mouth and dating from before 50,000 BP, probably as early as 60,000 BP according to Ralph Solecki, who led the excavations during the 1950s. Arlette Leroi-Gourhan, a palaeobotanist based in Paris, was responsible for analysing soil samples from the cave throughout the sixties and had found the job initially rather disappointing, as little pollen was contained in the earlier samples given to her for analysis and some levels of the site seemed entirely sterile. It was only in 1968, when samples with the numbers 313 and 314 from the grave of the Shanidar IV skeleton came to her for analysis, that something unique turned up. These two samples were unusual in that they contained clusters of more than a hundred grains of pollen, rather than the isolated grains that are typical of cave soils (where plants do not abound). Some of these pollen clusters actually preserved the form of the anther (the part of the stamen that holds the pollen) of the flower, showing that in these cases the flower itself and not just the pollen had been in the actual cave. In sample 313 she even found a scale of a butterfly wing, the calling card of a butterfly that must have visited one of the flowers some 600 centuries ago. It became clear to her in the course of the analysis that flowers of at least seven species could be identified in the soil deposits in the Neanderthal grave and must have entered this part of the site at the same time as each other. She had ruled out the possibility that they could have entered the cave through natural agencies such as the activities of rodents or birds, or through the coprolites (fossilised faecal matter) of mammals, and so was led to the conclusion that the flowers had been put there deliberately by Neanderthals for some symbolic purpose.

Others have sought more or less credible alternatives to account for the presence of the flowers. Gargett has suggested that the most obvious explanation is that the wind carried the flowers through the large opening of the cave, or that rodents might have brought them in whilst building their nests in the cave (there are fossil rodent holes in the cave). Leroi-Gourhan had already rejected the rodent option, and Solecki has been dismissive of Gargett's other proposed alternative:

> It would have taken a hurricane to blow bouquets of flowers (not single pollens) and very accurately to pinpoint the site of Shanidar IV. Moreover, the plants represented were in flower between the end of May and the beginning of July according to Leroi-Gourhan, and Shanidar IV was excavated on August 6.

The archaeologist Clive Gamble sees Gargett's theory of wind-borne flowers as a distinct possibility and also hints that the activities of local men employed to work at the archaeological site may be a factor, noting that Solecki himself reports the fact that one of his workmen placed yellow narcissi in the hollow handles of his wheelbarrow. Yellow narcissi do not appear among the flowers that Leroi-Gourhan has identified from the pollen found in the burial site. Yuri Smirnov accepts that the flowering plants were deposited by Neanderthals, but sees them as more likely to be bedding or covering of some kind rather than grave offerings; he also describes the presence among them of *Ephedra* as 'remarkable' (for *Ephedra* see below, p. 220). Despite the doubts raised and the alternative scenarios suggested for the presence of so many flowers in the cave burial, both Solecki and Leroi-Gourhan still remain convinced that their initial view that it was Neanderthals rather than winds or rodents is correct. They did, however, change their interpretation of the meaning of the placement of flowers. Originally they believed it was largely an aesthetic act similar in essence to the laying of flowers at a grave today. Having considered the properties of the main six types of flowers that were identified, they later suggested that some medical knowledge of the plants may have influenced the selection of these particular flowers by the Neanderthals.

The main flowering plants evidenced by their abundant pollen remains are all known to have medicinal properties, not only in Western folk medicine but also in the local herbalism that is still practised and that has been reported in the publications of the Iraq Ministry of Agriculture. Those of the *Muscari* genus, commonly known as the grape hyacinth, have stimulant and diuretic effects, although the bulb is toxic. The hollyhock (*Althea* genus) has been described by Solecki as 'the poor man's aspirin' on account of its use to allay pain caused by toothache and inflammation. Species of the *Senecio* genus (ragwort or groundsel) have a number of medicinal applications, including relieving female genital pains, and have also been used as haemostatic agents (i.e. as a means of preventing bleeding). Yarrow or milfoil (*Achillea*) has a number of uses in contemporary Iraq as an insect repellent, a means of treating dysentery and a general tonic. Species of *Centaurea* (St Barnaby's thistle) were eaten by locals during the time of the excavations at Shanidar and made into a number of herbal medicines.

The last of the six main plants is woody horsetail (*Ephedra*), which has a long history of use in Asian and other medical practices. It was once thought that *Ephedra* was the fabled *soma* of the ancient Indians, a psychoactive plant consumed by priests during their rituals. It is not a suitable candidate as it has amphetamine-like stimulant effects rather than the hallucinogenic properties attributed to *soma*. Nevertheless, it is known from archaeological sites in prehistoric Central Asia to have been consumed with more potent substances, such as opium and cannabis. Its more widespread use is as a remedy used to treat coughs and respiratory disorders, and in modern times extracts of it have been used to treat asthma. Another possible use that could have been made of *Ephedra* was as a stimulant to promote endurance and speed on protracted hunting expeditions and other tasks involving unusual levels of exertion. If the Neanderthals did indeed make use of these plants as a medical kit, then this would not really explain why they were laid on the grave of one of the members of a Neanderthal community. One could speculate that they may have believed that the healing properties of the plants might somehow aid the dead in the next world. Even more speculatively, it may be suggested that the dead individual had some connection with specific medical activities, perhaps even have been a medical practitioner.

Although such possibilities may seem far-fetched (and there is certainly no evidence that can really be said to support them), even those who are highly sceptical that the flowers were put in the grave intentionally by Neanderthals must consider the possibility of some basic medical knowledge at this remote time. For example, how is the apparently amputated arm of the Neanderthal man at Shanidar to be explained? Are we to suppose that the arm was simply severed with a stone tool and then left to heal on its own? This seems unlikely in the extreme and would in itself ascribe the Neanderthal body with rather extraordinary healing powers. To staunch the bleeding from such a wound would have required some kind of haemostatic agent. As we have seen, one of the plants found in the cave (*Senecio*) has just such a property, as does ochre. Amputation is in itself a surgical operation the outcome of which is not always as successful as it was in the case of the Shanidar man. Clearly amputation is not as difficult or delicate an operation to perform as the comparatively sophisticated surgery of Neolithic trepanation. In fact, a Neanderthal who was used to butchering animals on a regular basis would have had both the tools and the basic know-how to perform such an operation, particularly if, as may have been the case, the limb was partially frozen, making it both more insensitive to pain and easier to remove.

Although much remains to be discovered about the details of the Neanderthal diet, plants certainly played a role in their eating habits. It seems highly probable that their knowledge of the environment in which they dwelt (which we can infer by their very survival in it) included a knowledge of useful plants, useful not only for food but for health. As cats and many other animals are known to seek out plants with medicinal properties, it seems entirely reasonable to assume that the Neanderthal knowledge of such plants could have reached the stage of at least a primitive pharmacopoeia. The Neanderthal specialist Erik Trinkaus has pointed out an unusual practice that he was able to recognise as having been probably performed on two of the Shanidar skeletons (Shanidar 1 and 5). This practice is deliberate cranial deformation, which has been practised in more recent times for aesthetic reasons in places as far afield as New Britain in Melanesia and the Pacific Northwest coast of North America. It typically involves binding the head of an infant in order to alter the natural growth pattern of the skull to that of a shape deemed more pleasing. That the Neanderthals of Shanidar Cave may have performed cranial deformation is a surprisingly early manifestation of this cultural practice and shows a level of awareness that may indicate that they were not entirely lacking in anatomical and medical knowledge. Trinkaus sees its behavioural significance in the following terms:

This inferred presence of artificial cranial deformation among the Shanidar Neanderthals implies a heretofore poorly documented personal esthetic sense among these early humans. The appearance of this practice at the same time in human evolution as the first evidence of intentional burial of the dead and prolonged survival of the infirm would suggest a behavioral pattern allied with that of early anatomically modern humans.

Although the total number of known Neanderthal and other Middle Palaeolithic burials is rather small, attempts have been made to compare them with each other and with Upper Palaeolithic burials in order to see if significant patterns emerge. Burials are a prime source to archaeologists for providing information on the cultures to which they belong, and some investigators such as Yuri Smirnov and Francis Harrold have been able to gain some access to the Middle Palaeolithic world through the study of mortuary practices. Harrold has, for the purposes of his study, eliminated what he sees as dubious cases of intentional burial and bases his analysis on a total of 132 burials from Europe and Asia, 36 from the Middle Palaeolithic and 96 from the Upper Palaeolithic period. He recognises that the conclusions that can be drawn from such a survey, which covers a great geographical area and chronological period, are necessarily rather general.

The available sample of Middle Palaeolithic burials showed that overall more males than females were buried (the sex of many skeletal remains was not able to be determined), and this was particularly biased towards males in the western European cases (six males to one female), but since the sample was so small he did not think that too much significance could be placed on this ratio; in other words, he did not see this as evidence, in itself, of some preferential treatment of men or of a marked inequality in the status of the two sexes. However, when he looked at the data on the occurrence of grave goods (including bone and stone tools, animal bones and ochre), he found that whilst eight of the ten males were accompanied by such offerings, not one of the seven females was. This he believed was sufficient to imply that some higher status was accorded to males by the presence of grave goods and that this was likely to reflect social status in life. The burials of males were also found to be more diverse and more likely to show evidence of attendant features such as hearths and stone slabs which may have been part of funeral activities. Concerning age, Harrold was unable to detect any significant biases towards any particular group (for example adult males or children).

When looking at the Upper Palaeolithic evidence, Harrold found that the ratio of male to female burials was about 2:1 in favour of the former. He also discovered that differences in mortuary practices varied not only between regions during this period but also within a particular cultural complex. For example, he found that the most notable feature of the burials of the Magdalenian group 'is their variability; they include males and females, adults and juveniles, flexed and extended body position, and grave goods and associated features ranging from nothing at all to four categories of grave goods plus an associated hearth and stone structure. Comparable variability may also be found among other groupings.' According to Harrold, this may well reflect a greater variety of status roles in these societies compared with the simpler social organisations of the Middle Palaeolithic period.

Other comparisons made between the Middle and Upper Palaeolithic burials showed that the male-female ratio of burials was not significantly different between these periods, nor was the adult-child proportion (55 per cent adult in the Middle Palaeolithic, 63 per cent adult in the Upper Palaeolithic). Important differences were found in the prevalence of grave goods. As the great majority of Upper Palaeolithic burials were accompanied by grave goods, it follows that many of these – as opposed to none for the Middle Palaeolithic – were female burials. Perhaps the most striking difference between burials of the two periods (apart from the relative poverty of the former) was the occurrence of burials of more than one individual together – only a sixth of Mousterian burials were multiple whilst nearly half the Upper Palaeolithic burials contained more than one body (and, in the case of the site of Predmosti, Moravia, over twenty). It is interesting that these sites of multiple burial do not, in the main, seem to be like cemeteries, in which the dead are deposited over a period of time, but rather are the burial of numerous individuals at the same time. Harrold has speculated that these multiple deaths may have been due to disease, warfare or ritual human sacrifice, each or all of which can be shown to infer a higher population density and more complex social structure than that of the simple hunting groups of the Middle Palaeolithic. The overall picture gained from this review of the essential features of Harrold's comparative study of burials is one of an increase in diversity and complexity from the Middle to the Upper Palaeolithic, not only in the actual burials themselves but in the social world they are thought to reflect.

In many respects the research of Yuri Smirnov reaches similar conclusions to that of Harrold's analysis, although Smirnov makes it more explicit that the foundations of the human system of burial were laid during the Middle Palaeolithic:

Disposal of the dead is governed by certain laws, which we must understand if we are to explain the apparent anomaly that the first deliberate burials appeared as fully formed as Athena from the head of Zeus. Another important issue is that the forms of deliberate interment do not seem to have evolved. First, the positions in which the bodies were buried have been the same from ancient times to the present and almost all of the basic variations were already known in the Middle Paleolithic. Secondly, since the Middle Paleolithic all burial structures, whether natural or artificial, have been based upon two morphological archetypes, the pit and the mound (which, topologically, may be seen as the lower and upper halves of a sphere truncated by the surface of the earth). Third, grave goods, when present, are usually objects (manmade or not) used by the deceased during life (some things can be specially produced for interment) and belonging to no more than three categories – artifact manufacture, subsistence, and ideology; Mousterian burials exhibit an assortment of these ... Hence, the mortuary complex is a system consisting of two constants (a burial structure and human remains) and two variables (grave goods and associated features [e.g. hearths, pits, etc.]). Such complexes appeared as early as the Middle Paleolithic and have not evolved any fundamentally new features or designs since then.

Clearly from this point of view the cultural foundations of the human practice of intentionally burying the dead existed among both Neanderthals and 'Proto-Cro-Magnons' during the Middle Palaeolithic, and all subsequent styles of burial were essentially elaborated from this cultural inheritance. This places Neanderthals and their contemporaries firmly in the cultural realm and far from the animalistic lifestyle attributed to them in popular caricatures. For intentional burial necessarily indicates that a value was put on the dead; they were not merely discarded or forgotten but at least some of them left this world to some kind of ritual farewell from their family and community. Some of the burials still have the power to touch us now, particularly those of young children. At La Ferrassie in the Dordogne, France, skeletal remains of eight Neanderthals have been discovered by Denis Peyrony, Louis Capitan and others – the first seven (including the remains of a foetus) in the course of excavations between 1909 and 1920, and the other, a 23-month-old baby in 1973 by Henri Delporte. Although Gargett has tried to cast aspersions on the competence of the original excavators' ability to identify the difference between intentional burials and the workings of natural forces, his view has not gained wide acceptance. Although the excavations at La Ferrassie were conducted early this century and so without the aid of many archaeological techniques available today, this does not mean that the excavations were conducted in a slipshod fashion. Capitan and Peyrony were more than aware that the site was of great importance, and they actually delayed further excavations at one stage in order that other leading archaeologists could come to the site and witness the proceedings. Recent finds in the Near East have cast grave doubts on the opinions of Gargett rather than those of now deceased French archaeologists.

Grave goods have been found in the early burials at Qafzeh Cave and at Skhul Cave. At the first site a child burial contained an offering of deer antlers, and an adult burial from the second was accompanied by a boar jaw. Later burials in the region at Neanderthal sites reveal similar practices. At the Dederiyeh Cave in Syria, a two-year-old was buried with a flint, and at the Amud Cave near to the Sea of Galilee, a 10-month-old child had been interred in a recess with an offering of a red deer jaw bone. All these offerings to the dead – Neanderthal and modern types of human – have been discovered recently and by highly competent archaeological teams, and put any serious doubts that have been cast on the antiquity of burial practices to rest. If pre-Upper Palaeolithic peoples were expressing their sorrow at the passing away of their dead with funerary offerings and grave goods, then is it not also possible that they had other forms of symbolic expression that they may have exhibited in the form of art? It is this possible evidence for such activities that is explored in the next two chapters of the book.

Chapter 17

The Four Bones of Bilzingsleben

The controversy that surrounds the intentional human marking of animal bones is encapsulated in a one-page report by archaeozoologist (i.e. a specialist who identifies and studies animal remains from archaeological sites) Simon Davis which concerns six such objects found in Israel. Five of them came from an Aurignacian level at the Ha-Yonim Cave in the western Galilee region and are thus firmly dated to the Upper Palaeolithic period. The remaining one, from Kebara Cave, Mount Carmel, was firmly placed in the Mousterian layer below the Upper Palaeolithic material found above it; it is thus Middle Palaeolithic. The Upper Palaeolithic examples have between 3 and 15 notches on them and the Middle Palaeolithic one has about 14 notches. Davis, who has undoubtedly examined thousands of animal remains, is quite clear that in all of the six cases the notches are not the accidental by-product of removing meat or cutting tendons from the bone. Concerning the meaning that can be inferred from the incisions, he suggests that the Upper Palaeolithic specimens might be some kind of tally to record in a simple fashion the quantity of some object or thing that is unknown to us. Since the surviving fragment of the Middle Palaeolithic bone is clearly incomplete, having been broken at both ends, he states that he is unable to analyse the possible meaning that the object may have served. Yet, despite the smaller size and incomplete nature of the bone, the incisions are almost equal to the number found on the larger and later bone, and if anything display a slightly more complex arrangement. It follows from this that if we are to attribute the Upper Palaeolithic notched bones the function of a simple tally then why not the earlier Middle Palaeolithic piece?

The answer lies in the still widely accepted archaeological dogma that any form of symbolic activity – whether a tally or the making of a symbolic shape or art object – only begins with the Upper Palaeolithic period. This dogma has clouded the vision of many archaeologists in their investigation of such artefacts. When one looks at the sets of notches on the bones from the Middle and Upper Palaeolithic side by side it is clear that, if they are indeed intentional marks, then they must both be some simple kind of tally. Why is it that an example of a tally from the Aurignacian period is accepted by many archaeologists without batting an eyelid, but when a Mousterian example is proposed they instead raise their eyebrows? Such tunnel vision has nothing to do with scientific impartiality and it flies in the face of the evidence. It is as though prehistoric man had to watch the clock to see if it was the 'right' time for him to start making tallies and other such devices. Had the Mousterian notched bone been found in an Upper Palaeolithic context, then its status as a meaningful form of symbolism would have gone largely unchallenged; it is only its earlier date that makes it questionable to many observers.

Clearly the way forward is to concentrate on the evidence displayed on the object itself rather than be swayed by preconceived opinions which would have us deny the very possibility of its existence. Not all archaeologists have been so adamant in rejecting the possibility of such symbolism; as Marshack has pointed out, some early prehistorians entertained such thoughts. In 1909 Denis Peyrony had this to say about a bone discovered in a Neanderthal burial at La Ferassie, and clearly did not find the Middle to Upper Palaeolithic transition an impenetrable barrier to his free thought: '[It] contains a series of fine intentionally incised marks recalling the notched bones of the Aurignacian. Perhaps it had some meaning and was placed intentionally by the side of the corpse.' Marshack himself has made a microscopic analysis of an engraved bone fragment found by Janusz Kozlowski at Bacho Kiro, a Mousterian site in Bulgaria and estimated to be approximately 44,000 years old. He found that a zigzag motif on the bone had been made by the engraver without lifting his (or her) tool from the surface of the bone between the upward and downward stroke, thus showing that the making of a zigzag was intentional in nature rather than just the fortuitous meeting of marks caused by some utilitarian activity.

When the argument between the two archaeological camps (those who believed there was clear evidence for Middle Palaeolithic deliberate and non-utilitarian marking of bones and other objects, and those who sought other, utilitarian explanations for such phenomena) extended to a marked emphasis on the Lower Palaeolithic period some ten years ago, the debate reached a new level. For with this development the two camps found themselves proposing dates for the beginning of symbolic behaviour that differed from one another by over a quarter of a million years! This new phase in the debate was precipitated by the publication of findings from the Bilzingsleben Lower Palaeolithic site near Halle in former East Germany, dated to about 350,000 to 300,000 BP (although some archaeologists suggest it may be as late as 250,000 years ago, this does not really diminish the importance of the artefacts described below, since even this conservative date makes them extremely early). The site was discovered in 1969 and detailed excavations have been undertaken at the site for fifteen years by Dietrich and Ursula Mania.

According to these archaeologists, the prehistoric inhabitants of the site, whom they identify as *Homo erectus* (on the basis of analysis of seven molars and a number of skull fragments found at the site; some authorities consider them to belong not to *Homo erectus* but to an archaic form of *Homo sapiens*; I describe them as the former simply to avoid having to put both alternatives every time I mention the hominids that made the site), produced four bone artefacts showing undeniable signs of intentional engraved patterns. At this open-air site the Manias found remains of three circular or oval dwelling places, each of them three or four metres in diameter. These remains consisted mainly of piles of stones and bones which marked the boundaries of the dwellings; evidence for the existence of hearths was also found. Special tool-making areas which they call workshops were also found in front of the dwellings. They found that some of these primeval workshops still had anvil stones in place, and there was a great deal of stone, wood and bone debris that showed clear signs of human workmanship. At the time the site had been used by *Homo erectus* it was near a stream at the edge of a lake, and the archaeologists found not far from the main occupation area other zones used by the ancient inhabitants, including a rubbish dump. It was at the site of this ancient *Homo erectus* community that the four bone artefacts that have caused so much debate were found.

Artefact I was discovered between the workshops that were closest to the middle dwelling. It was made out of the tibia of an elephant, had at some stage been used as a tool and was broken at one end. As can be seen in *Figure 37*, the bone has a number of markings on it – a series of seven lines at the top and another fourteen lines in the centre. Analysis of the marks under the microscope showed them to be the same in diameter and to have been made by a single tool and most probably at a single sitting. For reasons which are not clear, the excavators considered that another sequence of seven marks was made on the lower part of the bone which had been broken. As Paul Bahn has pointed out, if there were a further seven lines below the surviving ones this would mean that the object would have a clear symmetry which would be unlikely to have occurred by chance. It is also worth noting that the extra lines would make the overall sequence 7-14-7, which aligns it with numbers seen to have special significance in the mathematical and astronomical notation systems described earlier in the book. However, such a numerical notation is extremely unlikely to have arisen at such an early date. Moreover the Manias provide no adequate explanation for proposing that another seven lines once adorned the object and so such speculations must be treated with extreme caution.

Artefact 2 (*Figure 38*) was discovered between the workshops outside one of the other dwellings and is part of a rib of an unidentified mammal. Both its ends had been cut off and the surface of the surviving part displays a number of roughly parallel lines. By looking at the four lines to the left it can be seen that they are actually each made up of three separate lines, and in each case the line linking the other two ends at the bottom on the right side (as we look) and its upper end touches the left side of the upper line. Measurement of the distances between the lines from left to right showed them to have a ratio of 3:1:2:2:3. As with Artefact 1, the lines seem to have been made by a single engraving tool. Artefact 3, made from a fragment of a

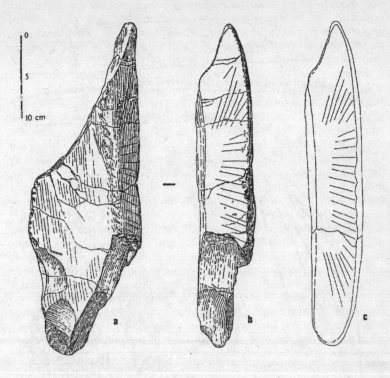

Figure 37

triangular elephant bone, was found in another workshop area (*Figure 39*). The five lines that can be seen extending from the top to the middle of the artefact are more strictly speaking five sets of double lines; perhaps, suggest the excavators, in order to make them more distinct. Once again analysis indicates the same tool was used to make all the lines on this artefact. Artefact 4 (*Figure 40*) was found at a wood- and bone-working area of the site and has a sequence of seven parallel lines which have a regular three-millimetre interval between them. Two of the seven lines (if counting from the lowest to the highest lines they are the first and the fifth) are over twice as long as the other five, and both of them are actually made up of two intersecting lines. The markings on this artefact were also made by a single implement.

So what, if any, meaning can be attributed to these modest but extremely old intentional markings? The excavators themselves have certainly not gone overboard in ascribing all kinds of complex and impossible-to-demonstrate meanings to the sequences of lines. For them, the site as a

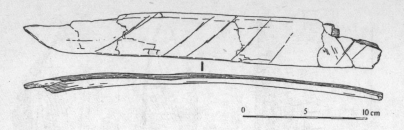

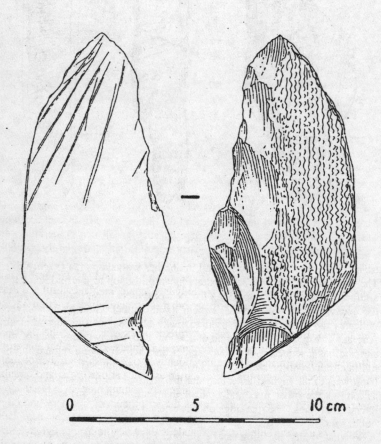

Figures 38 and 39

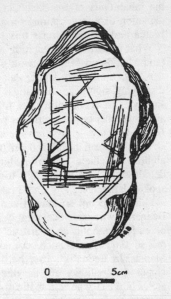

0   5 cm

0   5cm

Figures 40 and 41

whole gives a very good picture of the lifestyle of the *Homo erectus* community that lived there. They had dwellings, hearths, special areas put aside for industrial activities where a number of raw materials were fashioned into tools; they were clearly organised and sufficiently mentally developed to a level which would be unthinkable without at least a rudimentary language. With all these factors in mind, Mania and Mania suggest that the markings were not only intentional but meaningful and most probably had some mnemonic or other communicative function in the community in which they were made. They dismiss as too speculative the ideas of G. Behm-Blancke, one interpreter of the site, who believes that a bear cult was practised by *Homo erectus* at Bilzingsleben and that numerous other bone remains have been intentionally marked with a number of more complex decorative motifs, including crosses of various forms and circular and curved designs. These so-called signs are nothing more than the random marks caused by natural forces or as the by-product of butchery and other entirely utilitarian practices performed by *Homo erectus*. The bear cult, which was also attributed to the Neanderthals at one time, although this idea was subsequently shown to be based on the most meagre of evidence, is equally dubious, for although bear bones have been found at Bilzingsleben, there is nothing to suggest that any particular interest was paid to them by the inhabitants of the site. Claims have also been made that there is a depiction of a large animal on a bone object from Bilzingsleben, although Mania and Mania dismiss this suggestion.

What is particularly exciting about the four Bilzingsleben bone artefacts (and Mania and Mania have also found further such objects at the site) is that, unlike other pre-Upper Palaeolithic finds there is more than one object displaying significant markings, making them far more difficult to dismiss simply as freaks or anomalies. In fact, the arguments have been not so much about whether or not the marks were intentionally engraved on the bones by *Homo erectus* – this has been generally accepted – but about what significance can be placed on them and what they can tell us about the abilities of their makers. Mania and Mania believe that the marks demonstrate both a capacity for abstract thinking and the existence of a spoken language.

Others are not so sure. Even scholars such as Paul Bahn and Robert Bednarik, who have been highly instrumental in bringing together examples of early symbolic and 'artistic' objects in order to make archaeologists realise their significance, do not agree with all the speculations of the Manias. In Bednarik's case he has described resistance to the idea of pre-Upper Palaeolithic art as due to an enduring preconception among archaeologists who automatically reject such possibilities out of hand. He accepts the authenticity of the objects and does not believe that some utilitarian explanation (such as the use of the bones as cutting boards) will suffice in explaining how the marks got there. Concerning the third artefact and the five double lines on it, he suggests that a burin (engraving tool) with two projections was used to score the surface, thus producing two lines simultaneously, because 'it is absolutely impossible to produce such a perfectly spaced second line freehand'. He does not see the engraved bones from the site as evidence for abstract thought, but suggests it may indicate a stage of awareness towards the emergence of such a faculty; neither does he consider them evidence of the existence of language in this remote period.

A number of other bones from the Bilzingsleben site have been studied by Bednarik; one in particular is noteworthy. It is a bone incised with a much more complex arrangement of lines than the four bones mentioned so far (see *Figure 41*). At first glance it may seem to be just a muddle of lines that could have been caused during defleshing the bone, or even by natural forces. However, its shape makes it impractical for using as a cutting board and Bednarik makes the very apt point that if the lines were random and unintentional then they would have, at least in some cases, gone right to the edge of the available surface. However, as his illustration of the object makes clear, this has not occurred, and furthermore there is a certain symmetry to the overall pattern produced.

Bednarik notes that such artefacts are not entirely unique in the Lower Palaeolithic period and notes that under 500 kilometres away, at another Lower Palaeolithic site at Stránská skála in the Czech Republic, an engraved elephant bone was also found. François Bordes noted that such practices existed in these times when in 1969 he reported the discovery of an engraved rib (dated to about 300,000 BP) from the Acheulian layer of the Pech de l'Azé site in the Dordogne, France (the artificial nature of which has been brought into question), along with a Mousterian pierced object from the same site. Bahn believes that there are probably many more such artefacts from the Lower and Middle Palaeolithic period that have gone unnoticed by archaeologists, and that some such objects may be lying in museum collections where their true meaning remains unrecognised. Bednarik, despite his reservations concerning certain aspects of the interpretation of Mania and Mania, is in no doubt that the four bones of Bilzingsleben are of great significance:

> It emerges that the finds described by Mania and Mania represent the main body, and almost the sole unequivocal examples, of marked objects predating the Mousterian. They amply vindicate Marshack's and Bordes's courageous pronouncements regarding the Pech de l'Azé object, and they suggest that it may be time to seriously consider the existence of pre-Aurignacian rock art as proposed by Bahn and myself. The belief apparently held by most researchers, that art suddenly began with a big bang at the dawn of the Upper Palaeolithic, has no logical or factual basis. The figurines contemporary with the first incipient iconic art of Europe are so extraordinarily sophisticated that they can only be the result of a long tradition.

The opinions expressed above concerning the four bones and similar finds elsewhere are those of one school of thought that includes Mania and Mania, Marshack, Bednarik and Bahn among its members. A prominent proponent of the opposing point of view is Iain Davidson, who has been involved in many academic duels fought out on the pages of numerous learned journals during the last few years, particularly with Bednarik. Davidson accepts that the discovery of four engraved bones from the same site is not without significance and that the nature of the lines on them 'is suggestive that the creatures [note that their makers are not called humans] which [not "who"] made the lines were doing something more than accidental'. He is, however, highly sceptical of the Manias' overall portrayal of Bilzingsleben as a site that shows clear evidence of dwellings, workshops, big-game hunting and the like. He sees this as a fanciful interpretation of the *Homo erectus* lifestyle, which he believes can be reconstructed along more valid lines to show a far more primitive level of activity than that put forward by Mania and Mania. In stating his position, he invokes the authority of the grand master of his school of thought, Lewis Binford, who, he says, 'has devoted much effort to demonstrating the fragility of such myths'. Although Binford himself has shown no particular interest in art he has basically done more than anyone else to promote the view that there is no tangible evidence for real cultural activities before the onset of the Upper Palaeolithic period.

Thus both Bednarik and Davidson have described each other's position as based on a myth. Davidson decided to see the site for himself and so spent a week there at the invitation of the Manias in August 1989. His study of the site confirmed the position he had expressed before visiting Bilzingsleben; he interpreted it not as a hunting camp of the type proposed by the Manias but as a watering hole near which animals had died, and suggested that *Homo erectus* had scavenged meat from the carcasses. When he came to examine the four bones he thought the most likely explanation was that the marks were made during the removing of the flesh that was originally on the bones.

As Davidson had travelled all the way from Australia to visit the site, he also took the opportunity to have a first-hand look at a number of other objects in European museums that had been cited by Marshack and others as evidence of symbolic behaviour before the Upper Palaeolithic. Having inspected a number of objects such as grooved or pierced bones and teeth of a variety of animals, he came to the conclusion that none showed clear evidence of their supposed status as symbolic or ornamental objects, and thought that such a view was nothing but a 'sorry catalogue of unwarranted wishful thinking'. This remark, was in part directed at Marshack, who had published some of these objects as indicating ornamental and other symbolic activities of early man, and the article in which it appeared was swiftly acknowledged by Marshack, who in his reply was dismissive of Davidson's opinions. He described Davidson's visit to Europe as 'rapid' and 'touristic' and questioned the conclusions he was drawing based on a superficial study of the material by mere 'eyeballing' – Marshack's term for the study of objects without a microscope. He also thought Davidson was blinded by a doctrinaire position, writing:

> Unlike Davidson's study, my two-decade-long inquiry has not been conducted to prove either the existence of pre-Upper Palaeolithic symboling or to prove an entrenched hypothesis. It has been conducted to determine the nature of the problem and the supposed evidence, without preconception or a priori presumption.

The controversy over the Bilzingsleben bones and other early artefacts shows just how divided the archaeological community is over basic issues that are by no means resolved. It is not just an argument over a few marks on some broken animal bones; much more is at stake than that. It is about two fundamental and contradictory views of the origins of human culture itself, and of who was worthy of the name human. In one scenario, presently the most popular among archaeologists and which Andrew Duff and his colleagues call the Standard Model, there is an explosion of symbolic activity demonstrated by the earliest art and the emergence of what has been described as the Human Revolution; in short, the dramatic birth of culture as we understand it. Those who believe this is the true version of prehistoric events see the hominids that existed before this time as little more than animals or at best culturally impoverished creatures pursuing a miserable economy of scavenging rather than hunting. The occasional and awkward object that pops up from time to time from the earlier period and seems to be symbolic is dismissed as anomalous or insignificant.

In the other scenario, the so-called Cumulative Model, symbolic behaviour may be seen to have its origins in the Lower Palaeolithic period or at the very latest in the Middle Palaeolithic period. The small number of symbolic artefacts from these times is considered to be due in part to the simple fact that the further back one goes, the fewer objects will survive the ravages of time and the dramatic geological and climactic events that took place. Pre-Upper Palaeolithic hominids are seen as more like us in many respects and are often described as humans and attributed with at least a rudimentary culture and possibly a capacity for language. Although this paints the picture in a black-and-white fashion (and many archaeologists subscribe to positions of various states of grey, attributing certain of the above features to pre-Upper Palaeolithic hominids but not others), there are, as the debate over the Bilzingsleben bones shows, two schools of thought, one which sees the belief in pre-Upper Palaeolithic symbolic and artistic activity to be a myth and the other which sees the denial of this as based on a myth. So who is right? New evidence from excavations in the Golan Heights may be a decisive step towards the answering of this question.

Chapter 18

Graven Images from the Holy Land

The unusual honour of having found two particularly striking objects that point firmly to a pre-Upper Palaeolithic origin for art has fallen to Professor Naama Goren-Inbar of the Institute of Archaeology of the Hebrew University in Jerusalem. Both were found in the Golan Heights during the early 1980s and have been closely studied by Alexander Marshack. One of these two objects was found during the course of excavations in a 'demilitarised' area close to the village of Quneitra. It is a flat piece of flint (7.2 centimetres in length) that has been deliberately incised with four nested semicircles formed from a number of inter-linking straight and arced lines and accompanied by a number of subsidiary lines. This artefact is of Middle Palaeolithic origin and about 54,000 years old. Compared with other known Middle Palaeolithic engraved objects such as that from the Bacho Kiro site and from Kebara Cave, the Quneitra piece presents a considerably more complex design, although nevertheless rudimentary by our own standards. Unlike a number of other artefacts put forward as examples of intentional engraving that have been dismissed on the grounds that another plausible alternative for their existence can be found, it is much harder to argue that this design could be accidental or simply an amalgam of marks left as the result of some kind of practical activity. An overall design seems to be displayed by this object, particularly in the case of the nested lines, which were clearly planned.

According to Marshack, both the small size of the Quneitra flint and the design itself would require a considerable precision on the part of the engraver, and a complex co-ordination between the hand that held the flint and the one that held the burin or engraving tool. As he points out, this kind of manual dexterity was easily within the capabilities of both the Neanderthals and the anatomically modern humans who lived in the Levant at the time when the object was made (which of the two of them made the object is not known). Their known skill at making stone tools – attested to by countless numbers of such artefacts – is clear evidence for the existence of such capacities. No archaeologist has a problem with accepting that such skills were used for the practical purpose of making tools, but many still find it hard to accept that the same level of skill could be applied to the making of a symbolic object.

It is hard to explain this reluctance except as a result of preconceptions concerning Middle Palaeolithic humans – namely the entrenched idea that at this time they did not or could not create symbolic objects. Pieces of flint from later periods bearing both more complex and also simpler decoration than that found on the Quneitra flint have been accepted as symbolic patterns without hesitation. The 'unacceptable' nature of the Quneitra find is purely due to its early date, and it is clear that those who deny the Quneitra flint the status of a symbolic object are doing so on dogmatic grounds rather than on a purely rational basis.

Marshack thinks that flint was not the most common material used for making symbolic designs in the Middle Palaeolithic period but if anything was more likely to be an unusual choice. The excavation of the Quneitra site has shown it to be a seasonal camp and not a long-term base. Because of this, Marshack suggests that many materials that would have been typically used to make symbolic designs upon (such as wood, skins, etc.) were simply unavailable at this site, and thus flint was used out of necessity rather than choice. What the incisions on the Quneitra flint symbolise is impossible to say. Marshack's first impression on seeing it was that it looked like the arcs of a rainbow surrounded by rain; Goren-Inbar said the image seemed to evoke the volcanic landscape of the Golan Heights. Clearly hundreds of interpretations of the composition could be made, all equally subjective and impossible to relate to the intentions of its maker. Whilst the symbolic composition on the Quneitra flint has been found, the nature of that symbolism remains irrevocably lost.

During the course of excavations at the Acheulian site of Berekhat Ram in the summer of 1981, Goren-Inbar found another art object which is even more intriguing than the Quneitra flint. The object which has come to be known as the Berekhat Ram figurine is a small yellowish-brown pebble 3.5 centimetres long, 2.5 centimetres wide and 2.1 centimetres thick, and weighing just over 10 grams. Its natural shape is reminiscent of the female form, and this peculiar quality of the pebble has been made more pronounced by the addition of artificial grooves. More than 6,800 artefacts were found during the excavations, but the Berekhat Ram figurine is reported as the only art object among them. Its actual age is uncertain, but because it comes from a layer between two flows of basalt firmly dated to about 233,000 and 800,000 BP respectively, based on its presence in the Late Acheulian layer at the site, its minimum age cannot be much under a quarter of a million years old! The significance of this find cannot be overestimated, for this date makes it the earliest known figurine to be accepted by professional archaeologists.

As Goren-Inbar put it in her initial reporting of the object in an Israeli archaeological journal: 'the Acheulian figurine might be considered the earliest manifestation of a work of art, predating the Upper Palaeolithic art by hundreds of thousands of years'. That she announced this dramatic find in such a guarded way and in a very understated fashion was not because she was unaware of its profound significance, but rather because she was aware of its controversial nature. She was clearly waiting for confirmation that the grooves were definitely artificial and not the result of some unusual natural forces. Due to the fact that the figurine was a unique find at the site, and also because its human-like shape was natural, it was likely to be challenged or dismissed as a freak. Should it turn out that the grooves were not actually fashioned by early humans at all, then the object could be simply described as a manuport, the name given by archaeologists to a natural object collected by early man. Although this would still make the object an interesting and unusual find, it would be placed with finds such as the Makapansgat pebble (see *Plate XXVIII*), an object that has a natural resemblance to a face and was apparently collected for that reason by an Australopithecene. Since the Berekhat Ram pebble is so much later than the Makapansgat pebble, it would, if only a manuport, be no big news.

Its status as an art object was challenged by Andrew Pelcin of the University of Pennsylvania, who felt, after reading Goren-Inbar's report and looking at the illustrations of the object contained therein, that the grooves on the pebble could have been caused by geological forces. Based on Goren-Inbar's initial description of the object as a scoria pebble (scoria being the name given to certain material ejected from a volcano), Pelcin suggested that the grooves on the Berekhat Ram figurine bore a close resemblance to similar markings on other scoria pebbles that had been cast out from volcanoes, particularly those of a similar size that are known as lapilli (and sometimes by more colourful names such as cowdung bombs). Pelcin suggested that the status of the object could have been made clearer if other, indisputably natural, pieces of scoria from the Berekhat Ram site had been studied to see if they too had similar grooved patterns on their surface. If this was shown to be the case, then it was only the human-like shape of the Berekhat Ram figurine that would distinguish it from other scoria pebbles at the site. Furthermore, bearing in mind the small size of the object, it may not even have been noticed by the Acheulians among whose tools it was found; thus even its reduced status as a mere manuport would be spirited away. Although Pelcin's idea seems at least a plausible alternative, Goren-Inbar was clearly irritated by his suggestion and in response wrote:

> Reporting and describing finds recovered in excavations is the most fundamental obligation of an archaeologist; it is extremely annoying that individuals who never examined the figurine, and failed to read the detailed reports of the site and its finds, express their doubts concerning the excavators' observations and hence cast doubts on their abilities as archaeologists.

As it turned out, Goren-Inbar's original reporting of the grooves on the Berekhat Ram figurine as artificial was vindicated by microscopic analysis by Alexander Marshack and by a separate analysis of the object by Dr Sergiu Peltz, a vulcanologist (a specialist in the study of volcanoes) working for the Geological Survey of Israel. Peltz's examination of the pebble revealed that it was not a simple scoria but an amalgam of materials that included some pieces of scoria. He also was quite clear that some artificial modifications had been made to the object and so it could not be described as entirely a work of nature. Bearing in mind that not only did Peltz actually study the object himself but that he is also a specialist in the field with direct knowledge of the geology of Israel, his opinion carries considerable weight and effectively dismisses Pelcin's suggestion.

The right side of the figurine clearly shows how the head has been emphasised by the groove made to mark it off from the torso. The shoulder region has been artificially flattened and the right-hand flank of the torso, according to Marshack's meticulous analysis, appears to show an arm bent at the elbow. Large breasts clearly indicate that the figurine depicts a female. Unlike the right side, the left side of the figurine lacks an artificially worked shoulder, which Marshack says is due to an obstruction in the form of a small embedded pebble in this part of the object which would have made working the shoulder in the same fashion impractical. The lack of an arm on this side is due to similar obstructions. Nevertheless, even without a clearly defined shoulder or arm, the profile is unmistakably female.

When looked at from the front, the object shows a hole in the position one would expect the navel to be; however, there is no cleavage line to indicate the separation of the mass that has been identified from both profiles as the breasts. The central area of the front of the head is somewhat convex and has led Marshack to suggest this may have been the face. When this assumption is made, he feels that the surrounding area is suggestive of hair. Marshack's study of the back of the figurine indicates that there were two grooves made in the neck region – one coming from the front and the other from the back – and that they do not meet perfectly. This feature occurs on a number of later, and 'acceptable', modified pebbles from Israel. It will be clear from this description of the Berekhat Ram figurine that it has been produced by 'giving nature a helping hand' rather than being a work of art produced from scratch. The feminine form was fortuitously provided by nature and needed just a minimal amount of human modification to produce the desired effect. Considering its extremely early date, it is not surprising that it does not display all the features of a fully fledged artistic tradition. If one were to speculate on the origins of carving and sculpture, then it is not improbable that early examples would be modifications of naturally suggestive shapes and that the making of such works of art that were not so directly inspired by natural forms would have developed only subsequently.

The ability to make the grooves on even such a small object as the figurine was clearly within the technical capacities of the Acheulians of Berekhat Ram. A large number of stone artefacts found at the site – particularly those types known as end-scrapers and burins – show the precocious nature of their tool-making industry, and many of them are more reminiscent of the Upper Palaeolithic period than the Lower. Despite the fact that such tools from the site do not fit easily into preconceived notions of the technological level of Acheulian populations, unlike the figurine no one has doubted their authenticity. It seems, once again, that for many archaeologists there is a mental block concerning the acceptance of art from the Lower Palaeolithic period; the prevailing notion is that pre-Upper Palaeolithic man was a skilled tool-maker but could not produce works of a symbolic nature. The technical skill required to fashion the figurine was certainly not greater than that required to make the Berekhat Ram tools, so why is there such a resistance to denying Acheulian people the ability to produce such works? It is dogmatism and not reason that is the basis of this denial. In the light of the discovery of the Berekhat Ram figurine and the other suggestive evidence of symbolic capacities, there is no rational basis for assuming that the mental and manual capacities of Lower Palaeolithic people were only used in the production of objects of a strictly utilitarian nature. It is becoming increasingly obvious as the evidence mounts that Lower Palaeolithic hominids were far more human than many have claimed.

The precocious nature of some of the stone artefacts discovered at the site of Berekhat Ram, whilst unusual is by no means a unique occurrence. Other lithic assemblages that occur at the interface of the Lower and Middle Palaeolithic periods in the Levant also display levels of technological abilities not normally considered to have evolved until the Upper Palaeolithic period. Among such innovative industries are those known to archaeologists as the Amudian and the Yabrudian, which include various types of blade tools of an Upper Palaeolithic type. Middle Stone Age assemblages from southern Africa have also been shown to have foreshadowed later developments in stone tool technology. These industrial innovations, known as Howieson's Poort assemblages, are known from such sites as Klasies River Mouth, Boomplast Cave and Border Cave, and date from the period 90,000 to 60,000 BP. They contain Mesolithic features in abundance even though they are tens of thousands of years older than their Mesolithic counterparts. So how are such apparent anomalies to be explained; why do such comparatively complex technologies appear and then disappear again only to subsequently re-emerge in Upper Palaeolithic times?

Although precocious stone tool industries have been discovered and even recognised as such for more than 50 years, no satisfactory answers have been forthcoming to explain either how these innovations appeared so early or why they did not continue right through without a cultural break into the Upper Palaeolithic period. Although it is still not possible to answer these questions fully, there are certain indisputable facts which must be considered. The very existence of these artefacts at such an early stage shows beyond a shadow of a doubt that even as long ago as the end of the Lower Palaeolithic period early human tool-makers had the intellectual and manual capacities to produce tools normally considered to be only within the capacities of behaviourally modern humans, i.e. not before about 40,000 years ago. In an attempt to provide at least a partial explanation of industries which appear before they really should in the prehistoric technological sequence, L.B. Vishnyatsky of the Institute of the History of Material Culture in St Petersburg has stated that:

> Despite the absence of any generally accepted explanations, the very fact that there exists such a phenomenon as 'running ahead of time' in the development of Palaeolithic industries can tell us something important about the nature of cultural development in prehistory. Indeed, whatever were the causes of anticipation occurrences in different regions, they could not entail the consequences they did in the absence of one necessary condition which must be common for all these cases. To change substantially and rapidly, a culture must already have a great potential, a reserve of ideas and abilities which are known but not put into practice. Like the recessive genes in biology these ideas and abilities began to be materialized (to move from the genotype of a culture into its phenotype) only when circumstances (environmental, demographic, social, etc.) changed and the necessity arose to use the recessive part of cultural potential. Just as recent hunter-gatherers have not hurried to employ their extensive botanical knowledge in going over to agriculture, the bearers of Palaeolithic cultures used only that part of their potential of ideas and abilities which enabled them to lead a habitual way of life.

Thus, as hunter-gatherers in historical times have sometimes been reluctant to adopt an agricultural lifestyle, not because they were too ignorant or primitive to appreciate the importance of this supposedly evolutionary step but because they were simply happy to carry on as they were, so it seems that Palaeolithic peoples had capacities (reflected in their tool-making capabilities) that they chose not to continue using on a regular basis. Thus the return to simpler lithic technologies that occurred after the precocious beginnings of advanced kinds of tools can perhaps be seen as the lack of need or desire for the changes that such tools could make possible, and not just simply that such bold inventions were inadvertently forgotten. If tool-making skills intimately associated with the level of biological and cultural development ascribed to the Upper Palaeolithic period existed, at least in part, far earlier in prehistory, then it is quite possible that other cultural activities such as the fashioning of figurines could also have taken place much earlier than the 'acceptable' date for the origins of art at the dawn of the Upper Palaeolithic period.

I have shown throughout this book that the widely accepted view of the human story is wildly inaccurate. Many of the so-called innovations of 'civilisation' have been proved to be of Neolithic origin; much once attributed to the Neolithic Revolution is clearly Upper Palaeolithic; 'innovations' and advancements of the Late Upper Palaeolithic are shown to have existed in the Early Upper Palaeolithic; and the Human Revolution some 40,000 years ago – the purported origin of art, religion, language and even culture itself – has its cultural roots deep in the Middle and even Lower Palaeolithic periods. Early man may not have had all the capacities of we modern humans, but to deny him language, art, religious sentiments and other cultural achievements is to resist the evidence. Bearing in mind all that has been said about the capacities of primordial peoples, I will now turn to the question of whether or not some parts of the world were in fact colonised in a similarly precocious fashion – that is to say, much earlier than most archaeologists would accept.

Chapter 19

Dawn Stones or False Dawn?

To the outsider the study of the Lower Palaeolithic period may appear to be the least exciting of all archaeological pursuits, for whilst it is easy to see the romance and the attraction of digging at Ancient Egyptian, Mesoamerican and Mesopotamian sites, the comparatively modest finds from the earliest Stone Age sites seem to pale into insignificance in comparison with their illustrious descendants. The archaeologists who are dedicated to understanding the Lower Palaeolithic will seldom find more than stone tools and perhaps a number of animal bones showing the signs of human butchery; if they are very fortunate a few skeletal remains of hominids may also be unearthed. The stone artefacts which are the staple intellectual diet of prehistorians attract little attention when displayed in museums; their drab and dismal appearance may seem to epitomise the antithesis of the glamorous objects of later times such as golden masks, arcane inscriptions and monumental sculptures. Stone Age artefacts, particularly those of the earliest period, the Lower Palaeolithic, often look like a pile of old rocks, and the layman will find it hard to see the difference between such tools and the products of nature. This is hardly surprising, as archaeologists themselves often disagree about which are the work of humans and which the works of nature.

In the nineteenth and early twentieth century, heated debates among antiquarians and early prehistorians raged over the question of what was an artefact and what was not. Certain stones were called eoliths (from the Greek meaning 'dawn stones') on account of the belief that they represented the earliest kind of artefact that pre-dated even the Palaeolithic period. These objects were finally rejected and seen as geofacts (i.e. products of nature alone) rather than the work of primordial man. Nevertheless, even today, when archaeologists present what they believe to be stone tools of an anomalous age and crude workmanship, they are often dismissed as eoliths. The fascination for these earliest artefacts is partly due to the fact that they *are* the earliest; they represent the most primeval stage of material culture and often provide the evidence for the earliest peopling of the various parts of the world. They also present particular challenges in that their minimal nature requires the archaeologist to exert his or her ingenuity to the maximum in order to make them yield their secrets.

There are two basic questions that must be asked in assessing claims for tools which are controversial. Firstly, it must be established whether there is sufficient evidence to identify them as artefacts rather than geofacts by the way they have been flaked; corroborating evidence in the form of butchered animal bones or, even better, actual hominid remains, whilst highly desirable is rarely available. The second question is the dating of such finds, for even if they are shown definitely to be tools, how accurate are the claims of great antiquity that have been made for them? How these kind of questions are dealt with is best shown by describing some of the most controversial sites, which if shown to be genuine would shed an entirely new light on the capacities of our remote ancestors and the dates for the initial peopling of Asia, Australia and the Americas.

The hand-axe is the characteristic type of stone tool of the Acheulian industry of the Lower Palaeolithic period in Europe, Africa and Asia. Until comparatively recently it was thought to be a kind of tool that was not found east of India. However, the discovery of hand-axes in central Asia, China, Japan and Siberia has meant that this belief has had to be qualified, although most of these eastern hand-axes are not of the 'classic' forms known from India and further to the west. Despite its name, the hand-axe is not really an axe at all but most probably a tool used for a number of tasks. Many prehistoric archaeologists prefer to call these tools by the more neutral name of bifaces, on account of the fact that they are manufactured by working on both faces. The discovery of hand-axes and associated tools in India goes back to the 1860s, and numerous finds make the occupation of the region by early man uncontroversial. The earliest such artefacts found so far in India come from the site near the village of Bori in Junnar Taluka in the Pune District, and are about 670,000 years old. British archaeologist Robin Dennell has found hand-axes and other tools dating from the period 500,000 to 700,000 BP in northern Pakistan, but other discoveries made in the region by Dennell and his colleagues have raised more than a few eyebrows in the archaeological community. For near Riwat, south of Rawalpindi, far cruder kinds of tools were found which he claims may be about two million years old, and no less than 1.9 million years old. Although this would treble the time that early man is thought to have been in this area – a proposition which, in itself, is controversial enough – it also throws into disarray the whole accepted chronology for the movement of early man out of the African homeland.

Homo erectus is universally considered to be the first hominid to have left Africa, and according to current dating, not only did he not have time to race across to Pakistan to make these tools, he also, at least according to some archaeologists, had not even come into existence by this time, for his ancestor *Homo habilis* is only thought to have evolved into *Homo erectus* about 1.6 million years ago! A number of recent finds suggest that *Homo erectus* (or *Homo ergaster*, as some now prefer to call the African kind of *Homo erectus* who may be the ancestor of both a later strand of *Homo erectus* that moved east into Asia and also the earliest inhabitants of Europe) was on the move out of Africa considerably earlier than was previously thought. Acheulian artefacts from the site of Ubeidiya in the Jordan Valley, Israel, date to about 1.4 to 1.3 million years ago, and this dating is the earliest generally accepted evidence for the presence of any hominid out of the African homeland. The discovery of a human mandible at the site of Dmanisi in Georgia has been somewhat more controversially dated to 1.8 million years, although some would argue it may be just half as old as that. A date of about 900,000 years old would be more 'acceptable' because it fits more neatly with the overall chronology for the dispersion of hominids across Asia. Although it is still generally believed that most of Europe was unoccupied by hominids before about 500,000 years ago, recent evidence from the site of Orce in Andalusia suggests that at least parts of the Iberian peninsula may have been occupied as early as 1.7 or 1.8 million years ago.

But if *Homo erectus* was not the maker of the tools, then who was? His precursor *Homo habilis* seems an extremely unlikely candidate to have made such an epic journey, for whilst a tool-maker, evidence for the existence of this species beyond Africa simply does not exist. If the artefacts and their dates are accepted, then there are a number of equally controversial explanations. If the maker was *Homo erectus* then his presence in Asia would be contemporary with that of *Homo habilis* in Africa. If the artefacts were assigned to *Homo habilis*, who has been portrayed as a home-loving hominid, then he would have to be thought of as having reached not only beyond the African homeland but far to the east, at least as far as Pakistan. If neither *Homo habilis* nor *Homo erectus* made the tools, then there must be another hominid at work in this region during an extremely remote period. Bearing all this in mind, it is hardly surprising that the reports of two-million-year-old artefacts from Riwat have been looked at in a very sceptical light by most archaeologists, who have challenged both their status as tools at all and also whether the means by which they have been dated hold water. The artefacts from Riwat are made of quartzite and are very crudely manufactured (even widely accepted African artefacts from such an early time are very rudimentary in form), but according to Dennell they are nevertheless bona-fide artefacts. The difficulties in identifying them as such are noted by Dennell himself, who says:

It is often difficult to decide whether a lithic assemblage is the result of hominid (or human) flaking or of geological agencies. This is not a new problem in Lower Palaeolithic archaeology ... flaking to a repeated and regular format, for example, is a valid criterion if the artefact in question is a type that was made in a repeated and regular manner, for example, a handaxe ... it is far less useful when it comes to recognising waste flakes [i.e. the debris made as a by-product of stone tool-making activities] or the irregular type of flaking that characterises the earliest accepted (and many later) stone assemblages. The visibility of early hominid stone technology is, by nature, contentious: we can recognise as archaeological only those pieces of struck stone that are demonstrably different from geological flaking. Thus a cobble struck on one plane only and retaining most of its cortex [i.e. the crust or outer layer on a nodule of flint or other stone] is likely to have resulted from geological flaking; one struck in two planes and with a smaller amount of cortex is more equivocal; and one flaked over most of its surface from at least three directions may with a reasonable degree of confidence be attributed to hominid flaking.

Bearing such features in mind, Dennell and his associates rated the various possible artefacts from Riwat on a scale of 1 to 5. Those stones that scored only 1 were seen as objects that could most clearly be explained as produced by geological agencies; at the other end of the scale, a score of 5 indicated that the stone in question was clearly and convincingly an artefact. Scores in between were given to indicate the various degrees of probable and possible artefacts. Only one of the stones scored an outright 5, and is known by the number R014. R014 has a very large flake detached from one side and a further seven flakes had been detached from it. Two other stones, R001 and R008, received a rating of 4/5. R001 had been flaked in three clearly different directions, and 65 per cent of the original cortex of the stone had been removed during the flaking procedure. R008 not only showed signs that it had been worked on, but also 85 per cent of the cortex was absent.

The distinguished prehistoric archaeologist J. Desmond Clark (who, some 40 years ago, showed that the so-called Kafuan 'pebble tool industry' of East Africa that had been identified as consisting of man-made tools was actually nothing more than naturally occurring stones that merely resembled artefacts) has examined some of the Riwat stones and has expressed the opinion that they are indeed artefacts, although he believes them to be only about 500,000 years old rather than anything approaching two million years. If Clark is right concerning their age, then they would fit well within the generally acceptable range for artefacts from the Indian subcontinent and their highly controversial nature would therefore evaporate. Despite Clark's questioning of the chronology of these artefacts, Dennell has stuck to his guns by continuing to claim that they are as old as he initially proposed. Critics have attacked Dennell and his colleagues for not being rigorous enough in the criteria they used to distinguish artefacts from geofacts, and also for the lack of supporting data in the way of associated hominid remains or other indications of the presence of early man, such as butchered bones. But as Dennell has rightly pointed out, the tools found at the now famous East African site of Olduvai Gorge by Louis Leakey in the 1930s were initially identified only on the basis of the evidence of the tools themselves. In other words, the artefacts themselves were able to be distinguished from the geofacts which surrounded them by their own inherent characteristics and not by the discovery of hominid remains and other supporting evidence, which was only found *after* the initial identification of the Oldowan tools as genuine artefacts. Dennell also makes the point that numerous other tools of 'acceptable' dates found by archaeologists, including himself, have been accepted without the corroborating evidence of hominid remains or butchered bones being found among the debris at such sites.

Dennell's claims for a two-million-year-old occupation of Pakistan is by no means the only claim for the arrival of early man in Asia in such remote times. It has long been known that *Homo erectus* reached Java a million years or so ago, but it has recently been suggested that he may have arrived there as early as 1.8 million years ago. Although Java is not attached to the mainland of Asia any more, in the remote past it would not have been necessary for those who first arrived there to have crossed a body of water to do so. However, the discovery of artefacts on the island of Flores much further east along the Indonesian archipelago is a different matter altogether. The artefacts in question were discovered by Verhoeven in the 1950s in west central Flores and have been estimated to be somewhere between half a million and one million years old. Australian archaeologist Mike Morwood has re-examined the site and the artefacts from it and believes that the artefacts are of great antiquity. If this is the case then we have to ascribe greater capacities to the makers of these artefacts than most archaeologists would be willing to do. This is not because the artefacts themselves possess any precocious level of craftsmanship, but because to get to Flores from Java – even a million years ago – would have required the voyagers to cross 20-kilometre stretches of water.

To make such a journey would have meant that *Homo erectus* would have had to construct some kind of water transport (the simplest kind of craft would have been a raft) in order to get to Flores. This would not only mean that *Homo erectus* was in possession of much greater technological capacities than we thought; it would also be proof that language existed at this time, for no sea-going voyage is plausible without such a capacity. Whether Morwood's re-examination of these early tools from Flores will be accepted into the mainstream of archaeological thinking remains to be seen. If it is, then the capacities of early man will have been shown to be way beyond even what many advocates of the early development of cognitive faculties have dreamt possible.

At present the colonisation of Flores by sea remains highly controversial, and the generally accepted origin of water transport is assigned to a far later date. The evidence for the earliest water transport does not come from the discovery of remains of a boat or other craft, nor from the depiction of such a means of transport in rock art, but simply from the earliest date for the occupation of Australia, which could not have been reached by humans except by the crossing of water. A date of 40,000 BP is widely accepted for the initial occupation of the continent, although some archaeologists would place it somewhat earlier to 50,000 or even 60,000 years ago, as I will discuss below. There are early indications of the use of water transport from elsewhere in the world. The Aurignacian site at Fontana Nuova in Sicily dates from about 30,000 years ago, and it is generally thought that Sicily was not connected to mainland Italy at this time and so could only have been reached by the somewhat hazardous crossing of the Strait of Messina.

But to return to Australia, which still provides the earliest evidence for human voyagers, there have been a number of studies that have suggested that it was colonised by humans much earlier than 40,000 years ago. Discoveries at the Jinmium rock shelter, a site in the north-western part of the Northern Territory, hit the headlines in Australia and Britain when Richard Fullagar and his colleagues, who had been studying the archaeological finds there, announced that they had discovered the world's earliest art, said to be approximately 75,000 years old and thus about twice as old as the earliest generally accepted finds of art elsewhere in the world. They also claimed to have discovered stone artefacts that were over 100,000 years old and perhaps significantly older.

Among the stone artefacts recovered from the site was a quartzite object identified as a pounding stone due to the presence of starch residues on its surface which suggested that it was used in the processing of food, perhaps in the form of tubers. The evidence for art was in the form of cupules that had been pecked into fragments on sandstone (dated to a period before 58,000 BP), and the presence of ochre is identified as being connected with artistic activities at 75,000 BP. The estimated age of both the stone tools and the art work relied heavily on controversial methods of dating, and since the initial announcement of these finds from Jinmium the archaeologists involved have accepted that they seriously overestimated the antiquity of both the tools and the cupules. Ironically, no mention was made in the press (nor in the academic publications that announced the findings from this site to professional archaeologists) of the far earlier Berekhat Ram figurine when the Jinmium site was announced. This has become doubly ironic as the dating of Jinmium has turned out to be seriously flawed, whilst the Berekhat Ram figurine – despite its far more contentious date – seems to be a much firmer case for the archaic occurrence of art.

As I have said, there have been other researches in recent years that have delved into the possibility that the first Australians arrived far earlier than 40,000 BP. In the early 1990s, A. Peter Kershaw of Monash University, Clayton, Victoria, argued that there was strong evidence that the human occupation of Australia began as far back as 140,000 years ago. He claimed that the evidence for this comes from information obtained from two cores that had been drilled into the ocean floor off the coast of Australia to a depth of some 400 metres. The two cores are known collectively as Site 820. Such deep drilling allows scientists to gain information about the environment in remote times, and in this instance the base of the cores contained material from about 1.4 million years ago. Detailed study of the pollen and charcoal in the upper part of the core – covering the last 140,000 years – showed that there was a dramatic decline in the amount of pollen of certain plants which did not seem to be caused by climactic change nor by any other natural agencies.

The study of the charcoal levels found in the various samples taken from the cores revealed an equally dramatic increase, suggesting that fire was the major cause of changes in the pattern of vegetation. Kershaw believes that the large-scale burning of the landscape could best be explained by the presence of humans in Australia at 140,000 BP. Some of his critics have argued that the fires which caused such devastation could easily have been of natural origin and that nature would not have required a human hand to assist her in her work. Others have pointed out that if we are to add another 100,000 years on to the prehistory of Australia, where is the archaeological evidence to support Kershaw's case? For 100,000 years of occupation should result in numerous archaeological sites, at least some of which would surely have come to light by now. Despite the shortcomings of evidence from Jinmium and Site 820, proponents of an earlier occupation of the continent are still optimistic that more definitive data will be forthcoming. A couple of decades ago archaeologists believed that Australia had only been occupied by humans for a few thousand years and that the prehistory of the continent was a pretty short story. At that time a date of 40,000 years would have seemed like utter madness. If more substantial evidence of human occupation before that date comes to light then we may yet witness the discovery of an earlier period of Australian prehistory, one that may add a whole new chapter to the human story in this part of the world.

The antiquity of early man in China has long been recognised, and a date of around one million years for tools found in the Nihewan Basin west of Beijing is considered to be well established by archaeologists. Much earlier dates have been proposed; for example, artefacts from the site of Xihoudu in Ruicheng County, Shanxi Province, may be as early as 1.8 million years old. The artefacts include bifacial chopping tools and triangular points made from quartzite, but as Jia Lanpo has commented, although almost all the artefacts were found *in situ* they show signs of having been transported by water to the location in which they were discovered. If this was the case, then they may have been made at a much later date and their presence at the site does not necessarily indicate that they are as old as the date given for the site itself.

If we go east from China, we can get an insight into just how quickly the archaeology of an area can leap backwards in time, as has been the case in Japan. The first discovery of a Palaeolithic site in Japan took place just after World War II. Until this time so strong was the belief amongst archaeologists that there was no Palaeolithic at all in Japan that excavators of Jomon sites would stop digging once they had reached the bottom of the Jomon cultural layers, simply because the discovery of earlier artefacts was seen as totally improbable. Since the discovery of Palaeolithic sites all over Japan, archaeologists excavating Jomon sites have found that some do indeed contain Palaeolithic materials below these levels. Up to 1980 it was generally believed that there was no proper evidence for the human occupation of Japan before 30,000 BP. Although there were claims for the presence of quartzite tools of an earlier date, these were almost universally rejected as eoliths, i.e. geofacts. In 1980 artefacts from a number of sites, including Zazaragi, Miyagi Prefecture, Tohoku District (in north-east Japan), were reliably dated to before this 30,000-year-old watershed. Those from Zazaragi were dated to 130,000 BP although sceptics maintained that they were perhaps no older than 50,000 years.

By the mid-1980s other sites in the Sendai Plain in Miyagi Prefecture were pushing the dates even further back. The earliest artefacts from the site of Babadan were dated to about 200,000 BP. In the early 1990s artefacts from nearby sites continued this trend; the sites of Takamori and Kami-Takamori were found to be perhaps as old as 600,000 BP. Some of these very early artefacts were clearly demonstrated to be the work of early humans by the microwear analysis that was undertaken. It is thought that in this remote period of prehistory Japan was joined to the Asian mainland, and so these tools cannot be used as proof of any journeys made by sea. Thus we see how in the space of little more than a decade the acceptable dates for the human occupation of Japan went up half a million years! Things can indeed change rapidly in archaeology. It is not only in Japan, to the east of China, but also in Siberia, to the north, that there is growing evidence for the antiquity of early man. The belief that Siberia was occupied from a very early date is by no means a new idea; it has been among the proposed homelands of humankind since the nineteenth century. Vitaliy Larichev and other scholars at the Institute of History, Philology and Philosophy in Novosibirsk have described the early development of this notion:

With the exception of Africa, hardly another continent on earth has been the subject of such [a] persistent search for the cradle of mankind as has northern Asia. The rigorous spaces of the taiga, the permafrost zones, and the icy winds that blow off the northern ocean for most of the year should have precluded any thought that early man might have developed here. Nevertheless, paradoxical though it might seem, it was Siberia that occupied the minds of Western European scholars during the second half of the last century as they looked for the roots of the early European Stone Age cultures ... [This view] was keenly supported and further developed by Quatrefages, one of the most prominent French zoologists and anthropologists in the second half of the nineteenth century. He affirmed that Siberia, or the Far North in general, had been the cradle of mankind, where man had developed before the spread of the glaciations ... Unfortunately, the hypothesis of a northern Asian homeland for mankind had no basis in fact and was extremely speculative. It is merely one of the many imaginative reconstructions of the natural evolution of the world, which were typical of the natural sciences in the middle of the last century.

Due to the lack of evidence supporting this theory, its proponents became progressively few in number, and by the 1920s only a small number of diehards still clung to the idea. Hard evidence for the pre-Upper Palaeolithic occupation of Siberia has been a long time coming, and many are still sceptical of the ability of early man to survive – let alone thrive – in such a harsh environment. Despite the general reluctance to accept very early Siberian sites, some have been proposed.

Another scholar based at the same Institute in Novosibirsk was the late academician Okladnikov (who we have mentioned in Chapter 16 in connection with his discovery of the Neanderthal burial at Teshik-Tash, Uzbekistan) who, up until his death in November 1981, maintained that he had found clear evidence from a Lower Palaeolithic site that proved the antiquity of man in Siberia. The site of Ulalinka, in the Altai region of southern Siberia, was discovered in 1961, and its upper cultural level consisted of artefacts that were clearly Upper Palaeolithic in origin and were calculated to be under 25,000 years old. Below this Upper Palaeolithic level was what Okladnikov has described as the main cultural level. The artefacts found in this level were of a far more archaic type, which differed from anything seen in Siberia before. The simple tools were mostly crudely fashioned and split pebbles, mainly quartzite but with a few made from obsidian. Both their primitive nature and general peculiarity made them hard for Okladnikov to compare with tool types from other parts of the world. A number of dating techniques were used to try to establish their age. Okladnikov and his colleague G.A. Pospelova concluded that the artefacts must belong to the Lower Palaeolithic period and be more than 700,000 years old. Despite the fact that Okladnikov was the leading Palaeolithic archaeologist in the Soviet Union, the dating of the artefacts (and even their status as artefacts at all) was strongly questioned. Even the strongest supporters of the genuinely man-made nature of these artefacts have expressed doubts as to whether they could seriously be considered to be more than 200,000 years old. Most other archaeologists did not consider there to be sufficient evidence to substantiate the claim for artefacts at all, and the pre-Upper Palaeolithic finds from Ulalinka are now generally rejected. Nevertheless, the fact that the whole topic of the archaic occupation of the north had been raised from the dead did encourage others to continue the quest for very early artefacts.

In 1969 a series of early Palaeolithic sites were reported from the Angara basin in the middle Siberian Plateau and have been provisionally dated to around 200,000 BP, which the excavating archaeologist G.I. Medvedev considers a conservative estimate. The archaeologist Yuri Mochanov initially raised objections to these early dates but later changed his opinion and accepted their validity. In fact his own discoveries turned Mochanov's views of Siberian archaeology on their head. In 1983 he discovered the site of Diring-Ur'akh (see *Plate XXVII*) on the River Lena to the north of Yakutsk in Yakutia, and initial work there could hardly have prepared him for what he was about to discover. Excavations on the terrace of the River Lena had brought to light some human burials in stone coffins dating from about 1500 BC, and as the work progressed it became clear that these comparatively recent burials had been dug into Palaeolithic deposits. This was interesting enough, but when the antiquity of the earliest artefacts was calculated to some time between 1.5 and 2 million years, the site took on a whole new meaning, and it is no wonder that it has become known as the 'Pearl of Yakut Archaeology'!

Mochanov also found groups of natural pebbles at the site that seem to have been arranged artificially into circular formations, which could possibly be the foundations of dwellings made by the hominids who had also made the artefacts. Although the artefacts were of a very primitive kind, it was possible in some cases to fit back together the various pieces of pebbles that had been crudely fashioned into tools. This indicates that some of the tools found were actually made at the site and not just used there. Mochanov's analysis of the tools has led him to suggest that the oldest tools from Diring-Ur'akh are of a type only found elsewhere in connection with *Homo erectus*. It has also been suggested that they are more like those found in Africa that relate to the dawn of technology itself. Mochanov was converted to the notion of a very early occupation of the Far North, as he put it himself: 'I couldn't believe my eye, at first. After all, I had always argued against finding such primitive pebble tools in this part of Siberia.'

If the dating and the genuine status of the artefacts are shown to be correct, then the discovery of the 'Pearl' not only shows that Siberia was indeed a home to primordial man but also has other implications. Clearly this part of the world could not have been populated by early humans unless they had the ability to make (or at least preserve) fire and were sufficiently advanced to have made themselves fur clothing. We have already seen how contentious the issue of the Lower Palaeolithic domestication of fire is among archaeologists, and to claim this among the capacities of hominids even 1.5 million years ago is extremely controversial. Even if the site of Diring-Ur'akh is shown to be not nearly as old as Mochanov has suggested, the other Siberian sites that have been dated to some time before 200,000 BP still raise the same point concerning the requirement for fire and fur clothing. Most archaeologists simply do not believe that humans – even 200,000 years ago – were capable of colonising the harsh natural environment of the Far North because they were not advanced enough to control fire and make themselves clothes. But as Larichev and his colleagues have pointed out, the earliest people to enter Siberia cannot, in the light of the archaeological discoveries of Mochanov, Medvedev and others, be seen as 'instinctive, semi-animal beings' but rather as in possession of human minds. He states that 'The great significance of

Mochanov's discovery at Diring-Ur'akh is that it forces even the most confirmed sceptic to look again at the achievements of our early ancestors in the cultural and, especially, the intellectual spheres of their lives.'

Medvedev brings to the fore another possible implication of the discovery of early sites in Siberia when he proposes that 'The initial stage of development in the northern territories of Asia and America should not be very different in time, and, in principle, might go back to the Lower Paleolithic.' He is not the only archaeologist to express such an opinion. Thomas F. Lynch, in discussing the evidence for the early use of fire, has quoted researchers of the highly controversial site of Esperança Cave in Brazil (for which see the discussion of claims for the early occupation of the Americas below) as saying: 'It is of course not surprising that *Homo erectus*, who occupied the continent [*sic*] of China from at least 700,000 years ago ... and had domesticated fire 400,000 years ago (Zhoukoudien) would have crossed the Bering Strait a number of times.' It must be said that the idea of *Homo erectus* being the first American is, to almost all archaeologists, absolutely out of the question. The conventional view is that the Americas were first populated by Palaeo-Indians who entered North America from Asia 12,000 years ago, or perhaps a little earlier. At this time the two continents were not parted by the Bering Strait but were joined by the Bering land bridge. So it is something of an understatement to say that the idea of *Homo erectus* as the first pioneer of what would then truly have been the Wild West is outrageous indeed. Having seen that a number of archaeologists looking at things from north-eastern Asia have dared to propose the unthinkable, it remains to hear the dissenting voices in American archaeology that have also sought to bridge this primeval gap from the other side.

We have already seen that the Far North was once considered to be the cradle of mankind by some early investigators, and the idea that the original homeland was somewhere in the Americas also had a few proponents in these highly speculative times. One of the most extreme advocates of this school of thought was the Argentine Florentino Ameghino, who claimed (totally reversing the generally held view of the early movements and migrations of humankind) that Argentina was the place where humans first appeared and that they then travelled to North America and then onwards to Asia and Europe. To support his case, Ameghino claimed to have discovered very early stone artefacts in his country which he dated to between three and five million years old! He also reported his discovery of evidence for human remains, incised bones and the use of fire. His outrageous claims brought him into direct conflict with Ales Hrdlicka, one of the most influential figures in American physical anthropology and archaeology, who was sure that humans had only been in the Americas for a few thousand years. Ameghino's theories, like so many other early and unsubstantiated claims for human antiquity in various parts of the globe, soon fell into total obscurity, but others challenged Hrdlicka's very conservative estimates of the time scale of the occupation of the New World. Among these was Louis Leakey, the founder of the most famous 'dynasty' in the study of early man. Michael A. Cremo and Richard L. Thompson, in their compilation of renegade and anomalous archaeological theories and discoveries, entitled *Forbidden Archeology*, quote Leakey's recollection of his first conflict with Hrdlicka:

> Back in 1929–1930 when I was teaching students at the University of Cambridge, I began to look into the question of the antiquity of man in the Americas. Although there was no concrete evidence to indicate a remote age, I was so impressed by the circumstantial evidence that I began to tell my students that man must have been in the New World at least 15,000 years. I shall never forget when Ales Hrdlicka, that great man from the Smithsonian Institution, happened to be at Cambridge, and he was told by my professor (I was only a student supervisor) that Dr Leakey was telling students that man must have been in America 15,000 or more years ago. He burst into my rooms – he didn't even wait to shake hands – and said, 'Leakey, what's this I hear? Are you preaching heresy?' I said, 'No, sir!' Hrdlicka replied, 'You are! You are telling students that man was in America 15,000 years ago. What evidence have you?' I replied, 'No positive evidence. Purely circumstantial evidence. But with man from Alaska to Cape Horn, with many different languages and at least two civilisations, it is not possible that he was present only the few thousand years that you at present allow.'

Ironically the 'heresy' of Leakey is closer to the orthodoxy of today than Hrdlicka's estimations were at this time; in fact, artefacts found in the 1930s were demonstrated in the 1950s to be 12,000 years old. Whilst some modern archaeologists have viewed Hrdlicka as a meticulous and sceptical scholar who rightly resisted claims for an earlier occupation of the Americas because hard evidence for this was simply not available at the time, others have portrayed him as a dogmatist determined to defend his position at almost any cost, even by the suppressing of evidence from sites that did not fit with his beliefs. As in most such polarised arguments the truth is somewhere in the middle. Leakey's audacious statement in his early days has now been vindicated, but at this time even he could not know how radical a heresy he was to propagate in his twilight years, and that this heresy would also be concerned with the problem of the first occupation of the New World.

This latter heresy, unlike his earlier one, has not been vindicated from the point of view of almost all archaeologists. Leakey's work in East Africa had earned him the universal respect of archaeologists and anthropologists, and Olduvai Gorge became one of the most famous archaeological sites in the world. Whilst his wife Mary Leakey and their son Richard continued the work at Olduvai, Louis had his sights set on research elsewhere. According to Delta Willis' account of the life and works of the Leakeys, Louis had begun to feel envious of his wife's discoveries at Olduvai and her reputation in the scientific community, which had now outstripped his own. Willis also describes the intensely competitive nature of his relationship with his son Richard. What happened next is usually portrayed as an embarrassing end to Louis' career; that is, if it is even mentioned at all. Louis picked up his old interest in the question of the first peopling of the Americas by involving himself in excavations of the Calico Mountains site in the Mojave Desert, California, and in doing so, some would argue, was to be misled by eoliths in the twilight years of his career.

During the 1950s the San Bernardino County Museum undertook a large-scale archaeological survey. In the course of this investigation a considerable number of what were apparently artefacts were discovered at what was subsequently named the Calico Mountains site. Specimens of these artefacts were shown to a number of domestic and foreign experts by Ruth D. Simpson, including, in 1958, Louis Leakey. He was sufficiently interested to decide to make a visit to the site in 1963, when he was able to see artefacts still *in situ* at the site. The following year Leakey initiated excavations at the Calico site, believing that this was a major archaeological discovery and perhaps, in its own way, his New World equivalent of Olduvai. He was to continue working there, directing the excavations with Ruth Simpson, until his death in 1972. Leakey sifted through the numerous specimens that were discovered during digging at Master Pits I and II and expressed the opinion that more than 600 were clearly man-made artefacts.

He stated that 'The artifacts which have come out of these two pits completely satisfy me that we have at Calico a site with clear evidence that toolmaking man was here over 50,000 years ago' and: 'I say that some of the Calico specimens are not, and could not be, the work of nature . . . I say without hesitation, knowing what nature can do, that we are digging in an archaeological site.' And again: 'For me, after 50 years in the field, there is no doubt, whatsoever!' The San Bernardino County Museum team have, since these comments of Leakey, dated the artefacts from Calico to a far earlier time than even that proposed by him, namely to about 200,000 BP. Of the Calico specimens in question, about 80 per cent were made from chalcedony and the bulk of the rest are jasper. According to Simpson and her team, a number of distinct types of artefacts, including hand-axes, hammerstones and scrapers, were found at Calico, and their forms could not be the result of natural forces but can be nothing else but the tool kits of 200,000-year-old occupants of California. They also claim that these artefacts are of a comparable technological level to those found at Lower Palaeolithic sites in China, and see the lack of acceptance of their finds as indicating a psychological barrier on the part of most archaeologists in accepting new and controversial data that does not fit neatly into preconceived notions of the antiquity of humans in the Americas. A semicircular arrangement of stones found in Master Pit II has been interpreted as the remains of a hearth, but this has been treated with equal scepticism.

The status of the Calico finds highlights how even major figures such as Louis Leakey have become embroiled in the debates concerning the fundamental question of what the criteria are that can effectively distinguish the works of nature and of humans in remote prehistory. The majority of archaeologists reject the Calico specimens even though a senior figure like Leakey saw them as unquestionably artefacts. If we are to follow this widespread consensus then we still have to ask ourselves how it was that so experienced a field archaeologist was misled into believing them to be the genuine article, which, if nothing else, shows the scale of the difficulties in assessing such eoliths or dawn stones. The proponents of Calico and other very early sites would argue that despite the artefacts being recognised by such established authorities as Leakey, the pervasive dogmatism of American archaeology that denies the very possibility of such early sites has prevented an impartial assessment of the evidence. Opponents see the situation as a straightforward case – there is no concrete evidence that the eoliths of Calico are man-made at all, and as such the finds are no more than stumbling blocks for those who continue to try to demonstrate their status as artefacts.

Similar doubts have been expressed by the archaeological community at large concerning analogous claims from South America. Maria Beltrao, whilst excavating at Toca da Esperança ('Cave of Hope'), Bahia State, Brazil, during the 1980s, claimed to have found artefacts that are between 204,000 and 295,000 years old. Both the dating of these objects and their identification as humanly modified stones have been widely criticised. De Lumley, who has supported the validity of these Brazilian finds, has linked them with the Calico specimens in arguing that *Homo erectus* is the most likely manufacturer of such earlier tool kits and was thus at least intermittently able to enter the New World from Asia. It is not only the dating and the status of the proposed artefacts from both Calico and Esperança that make most detractors sceptical; it is also the complete lack of acceptable finds of human or hominid skeletal remains anywhere in the Americas that could be dated to anything even remotely close to these extremely early dates.

Unless new and more obviously convincing evidence comes to light, the occupation of the Americas before 200,000 years ago remains to be proved. So notorious are claims of this antiquity that many archaeologists who are seeking to establish comparatively modest claims for extending the period of human occupation of the New World have done much to distance their own work from these more radical claims. A growing number of archaeologists are accepting dates for the presence of humans in the Americas back to 15,000 or even 20,000 BP. But between such claims on the one hand and Calico and Esperança on the other lie mid-range sites, the most talked about of which is controversial but nevertheless not always perceived as belonging, chronologically and psychologically speaking, to 'the outer limits'.

The rock shelter known as Toca do Boqueirão da Pedra Furada (or more simply as Pedra Furada) in north-eastern Brazil is probably the site that has aroused the most controversy, as well as the most constructive debate, for whilst many archaeologists have expressed considerable reserve over the status of the finds there, they have, at least in some cases, kept an open mind. Brazilian archaeologist Nième Guidon has spent more than 20 years undertaking research in this semi-arid region of Brazil and has documented about 350 sites in the region, many of which display rock art. The site in question was discovered in 1973 and is a large painted rock shelter some 70 metres wide and, at its greatest extent, 18 metres deep. When she first started working at the site, Guidon did not expect to find evidence of human activity going back beyond the 'acceptable' upper limit of 12,000 BP.

Her initial interest was to try to find out how old the paintings at the site actually were. In order to accomplish this, she began excavating at the base of the decorated walls and found both stone artefacts and evidence for hearths in the form of charcoal deposits. The excavations began in 1978, and it was not until 1985 that Guidon and her team finally reached the bedrock some three metres below the sediments in which they had been digging. During the course of this work, samples of the charcoal found at various layers were sent off to France for carbon dating. The first two dates to be sent back to her were 7,640 ± 140 BP and 8,050 ± BP, and they were therefore not controversial. But as the subsequent samples were dated to increasingly early dates, the site began to take on a whole new significance, and it was its apparent age that became the central concern. The art in itself, interesting as it was, could simply not arouse the same attention as the accumulating evidence of what was rapidly becoming one of the strongest cases for the presence of humans in South America way before the 12,000 BP watershed. So the laboratory results came back with approximate dates of 17,000 BP, 25,000 BP, 32,000 BP and finally reaching back as far as 48,000 BP!

At first Guidon, rather than being elated, was shocked. This was something she was totally unprepared for, as she had never set out intentionally to find evidence of such a controversial nature. She had no ideological axe to grind, no dogma to prove. The dates seemed so incredible that she presumed there had been some error in the dating, but having checked this the dating stood firm. On the basis of the early dates given for the site, Guidon has suggested that archaeologists should stop thinking of the Bering land bridge as the only possible way into the New World from Asia during early prehistoric times. She suggests that entry into North America could have been achieved by a series of short sea journeys from island to island along, for example, the chain of the Aleutian Islands. Whilst this cannot be entirely ruled out, there is no evidence for a comparably early presence of humans, at least not in any of the archaeological sites known in this particular island chain. Leaving aside this speculative theory that encompasses regions far from Brazil, in order to see why Pedra Furada has caused such divisions amongst archaeologists it is necessary to describe it in more detail.

The site has two main cultural layers (each containing a number of internal divisions which need not concern us here). The upper layer is called Serra Talhada and is dated to after 10,400 BP, and as such has aroused little or no controversy. The lower layer, Pedra Furada, starting at around 14,300 BP, is controversial from top to bottom and it is for this reason that the finds from this layer of the site have been subjected to close scrutiny by sceptics. The evidence on which Guidon and her team have built their case consists of the radiocarbon dating of charcoal which they believe to be the residual evidence of human campfires, and the 600 or so tools made from quartzite found in the Pedra Furada layer. While most critics accept that the radiocarbon dates are reliable, they do question whether or not the charcoal which yielded these dates is really the result of human activities, and have suggested that it may be the result of bush fires instead. As is usual for sites for which highly anomalous dates have been claimed, the status of the artefacts has also been questioned, and the possibility has been raised that the Pedra Furada tools are nothing more than rocks that have fallen down from above the rock shelter and split in ways that may be mistaken for the handiwork of humans. Thus the sceptics have argued that whilst Serra Talhada may be a genuine archaeological site, the earlier phase of use of the rock shelter may be nothing more than an illusion created by falling rocks and bush fires.

Guidon is dismissive of her critics' alternative scenario for the Pedra Furada site and has pointed out that not only does she have decades of experience working in this kind of environment, but also that the majority of her critics have no first-hand knowledge of the site, and that if they did then some of their objections could be answered on the spot. As Paul Bahn has pointed out, the degree of scepticism displayed towards the site of Pedra Furada may well have been heightened by the fact that Guidon, as a woman and a Brazilian, has been seen as a less 'acceptable' kind of archaeologist than if she were a North American man. Whilst such prejudice has not been openly voiced, it cannot be ruled out as a factor in the argument over the status of the site.

Late in 1993, three leading archaeologists from North America (all men!) went to Pedra Furada to see the site for themselves, at the invitation of Guidon and her colleagues. All three had been involved in debates over the first peopling of the Americas for a number of years, making their visit all the more interesting for themselves as well as for Guidon's team and the archaeological world as a whole. Two of them, James M. Adovasio and Tom D. Dillehay, are well known for their own excavations of sites for which pre-12,000-year-old claims have been made. Adovasio's work at the Meadowcroft rock shelter in Pennsylvania is widely considered to provide the best evidence for the earlier occupation of the continent (at least 14,500 BP), whilst Dillehay's site of Monte Verde in Chile has produced strong evidence of human activities before 12,500 BP and perhaps even as early as 33,000 BP. In a report that they made after visiting the site, they pointed out that since they were both advocating their own sites as evidence of the pre-12,000 BP occupation of the Americas, they were hardly likely to be critical of the possibility that the site might have been as old as its excavator claimed, and therefore their report would be unbiased by any dogmatic preconceptions. The third visitor was David J. Meltzer, who has been a central player in the whole debate concerning the evidence for the early populating of the New World.

Like many of the critics who voiced their negative opinions from afar and did not have the opportunity (or inclination) to visit the site, Dillehay, Meltzer and Adovasio were unconvinced by the published evidence resulting from Guidon's excavations. The visit to Pedra Furada did not change their opinions in any significant way and they found neither the charcoal deposits nor the claimed artefacts convincing. They felt that nothing about the charcoal from the Pedra Furada cultural layer indicated that it was the result of human agency at all, and that it was simply evidence that there had been bush fires at the site in the past. They also compared the charcoal deposits in this layer unfavourably with the clearer evidence for hearths in the much later Serra Talhada phase and the evidence for the same phenomena in other cave and rock shelter sites that they had personally investigated. In 1996 Guidon and her fellow researchers went into print to refute some of the criticisms that had been levelled at them by their three distinguished visitors. For despite the experience of the visitors at other sites, Guidon argued that they had got it all wrong concerning Pedra Furada, pointing out that the surrounding area did not show comparable evidence of the natural fires claimed by Adovasio and the others. The point they were making was that if the Pedra Furada fires were natural, why on earth were there no signs of these 'natural' fires in the vicinity of the site? For Guidon, the idea that natural fires would have limited themselves to the site alone was ridiculous. But most of the ink was spilt over the status of the artefacts which for Guidon & Co. was the cornerstone of their evidence.

One way in which contentious claims for the existence of artefacts can sometimes be settled is by the process known as microwear analysis. This kind of procedure was first developed by the Russian researcher Semenov in the 1930s and later refined by the American archaeologist Larry Keeley. I have mentioned microwear analysis in passing elsewhere in this book, but I will briefly describe it again here, concerning its role in the Pedra Furada case specifically. Keeley distinguished distinctive kinds of polish on prehistoric stone tools which could only be seen under the microscope. Simply put, each type of polish indicated that the tool in question had come into contact with a particular material (e.g. wood or leather), and it was therefore possible to discover what a tool had actually been used for. Thus the study of an artefact under the microscope can reveal evidence of damage caused to the edge of tools. The technique is sometimes useful in assessing the status of a dubious artefact, because if the stone in question has traces of microwear then this clearly shows that it is indeed the genuine article.

Unfortunately such tests cannot be applied across the board with straightforward and unambiguous answers. This state of affairs is not only due to the fact that a real tool could have been made but never used (and thus would show no signs of microwear despite its authenticity), but also because whilst microwear studies of flint tools have been very useful in providing information which would otherwise remain entirely unknown, the application of the technique to tools made from other types of stone has not been so clear cut. Quartzite, unlike flint, is not generally as accommodating to the microwear analyst, and this has sometimes limited the usefulness of the technique, not only in the case of the materials from Pedra Furada but also with many of the other controversial artefacts mentioned above which are also of quartzite, such as those from Pakistan and Siberia (although as I have said, microwear traces have been found on Palaeolithic quartzite tools from Japan).

During his stay at Pedra Furada, Tom Dillehay did make a cursory microwear analysis of a few of the artefacts, concentrating on chopper-type tools which one would presume to have had heavy use and thus greater (and more easily detectable) use damage. The analysis found no evidence of use wear, although Dillehay did say that the small number of purported tools he studied in this way was insufficient to reject out of hand the possibility that others from the site might show such traces if they were examined under the microscope. He conceded that natural agencies such as chemical wear and water percolation could have worn away the evidence of the use of these tools during the extremely long period since they were made. Dillehay also conducted a small-scale project of experimental archaeology to augment the negative conclusions of his analysis of the claimed artefacts. He took a few samples of sharp-edged quartzite at the site and made them into experimental replicas of simple tools by subjecting them to about 200 chopping and 400 cutting strokes. These replicas were then examined under the microscope. They showed signs of damage – something that had not been found on the reported artefacts. Guidon and her associates were unimpressed by this part of Dillehay's on-site assessment of their site. After noting the complex and often obscure nature of results gleaned from microwear analysis, they stated that his examination was too cursory and hinted that higher magnification than that used by Dillehay (50 x) may have picked up traces which he was not able to see.

But it was not the microwear analysis that was the visitor's main reason for rejecting the artefacts (for, as I have said, they felt it to be inconclusive even from their own perspective), but rather their belief that natural causes could explain these eoliths away. They thought that the most likely explanation for all the flaked stones that were alleged to be artefacts was that they were geofacts made from cobbles of quartzite eroding out of the gravel bar about 100 metres above the rock shelter, which is linked directly to the site by the chutes at both ends of the shelter. In their opinion, 'These chutes have been and are veritable geofact factories.' In their reconstruction of events, the eroding cobbles fell the 100 metre distance and broke on impact with the floor of the rock shelter. The flaking or fracturing of these cobbles on striking the ground was what created their form – it was the work of nature not of humans that made the artefacts. One of Guidon's co-workers, Fabio Parenti, had already conducted his own experiments in order to determine if such a natural explanation was as plausible as it might appear. He examined 2,000 naturally occurring stones from the base of the chutes and found that not even a single one of them had fractured or flaked in a pattern that resembled the artefacts discovered by the team at the site. For him this showed that the idea that he and his colleagues had been duped by the fortuitous falling of rocks was utterly unfounded.

Robin Dennell had had to deal with a similar possibility with the quartzite tools that he had discovered in Pakistan. In order to demonstrate that the Riwat artefacts were not made by pieces of quartzite smashing on the ground or against other stones, he conducted an experiment which he and his colleague Linda Hurcombe describe as follows:

> The experiment is as follows, should anyone wish to replicate it: as principal investigator (PI) take 100 quartzite cobbles/unflaked stone nodules that are 5–15 cm (3"–6") long (i.e. the size that could have been used for making stone tools) one concrete, steeply-sloping embankment around 12–15 m high; and (indispensable) one gullible collaborator (GC). Place the GC at the bottom of the embankment to collect the resulting fractured stones. Start by throwing each stone as high in the air as possible, so that it strikes the embankment at least once when it falls. As the experiment progresses (and no fractures occur), proceed by throwing the stones as hard as possible on to the concrete embankment. Advise (persuasively) the GC to take cover as the stones are hurled down, but tell him/her to be absolutely sure to retrieve each stone as it whizzes by. End (in complete frustration) by offering to change places with the GC to show that no personal feelings are involved, and see if the GC can induce any flaking in this manner!

Although they portray the experiment in a humorous fashion, the point is, nevertheless, serious. For it showed that tool-like fractured stones were simply not produced by throwing the stones from a considerable height on to a hard surface. They therefore concluded that the falling of pieces of quartzite down a cliff or a chute would not produce flaked and fractured stones that were reminiscent of artefacts, and they felt this was true not only in the case of the Riwat artefacts from Pakistan but also in that of those from Pedra Furada. They do concede the possibility that if they had used a larger sample of stones (thousands rather than only a hundred) then some geofacts that looked like artefacts may have been formed, but they believe this to be a possibility 'as remote as the proverbial monkey typing Shakespeare'.

Having reviewed a number of contentious sites across Asia and America, we have followed a route that has taken us from Pakistan all the way to Brazil. In each case there has been a fracturing of the archaeological community between those who accept some of the finds and their early date (such as Dennell, Leakey and Mochanov) and others who reject the idea that genuine artefacts have been found at these sites at all. Among the latter is Nicholas Toth, Co-Director of CRAFT (the Center for Research into the Anthropological Foundations of Technology), based at the University of Indiana, who has surveyed a number of the various claims for early stone artefacts, including those from Calico, some of the South American sites and also Europe, and found them lacking sufficient features categorically to accept them as artefacts. Yet concerning the artefacts of Lower Palaeolithic Africa, an area of the world in which very early dates are more likely to be acceptable to most archaeologists, he and his colleague Kathy Schick believe that it is likely that there exists an earlier period of tool-use which we cannot detect because such tools were so crudely made that at present, and possibly forever, we are unable to demonstrate that they were unequivocally artificially made.

So, are the stones said by their excavators and other supporters to show a human presence in remote times in places as far afield as Flores, Siberia and South America merely phantoms conjured up by prehistorians from the rubble of nature, or are they the stepping stones by which the primeval journeys across the seas and arctic deserts that were made by *Homo erectus* can be discerned? Are they simply the product of archaeological flights of imagination, or are they milestones of sojourns that represent an intrepid spirit of exploration by our much-maligned ancestor at the dawn of time? It is at this point that culture and nature meet face to face, where the evidence for technology and for culture must be accepted or rejected on the basis of flaking angles and other features of these dawn stones. Here the flickering light of culture fades imperceptibly back into nature and into a dark and stony silence.

Afterword

In this course of this book, the human story has been traced from the cusp of history back to the inscrutable nature of the earliest known artefacts, which, in the final analysis, are barely distinguishable from the works of nature. This journey back to our cultural origins has unearthed some surprising results. That cultural activities such as the making of oriental rugs, the use of the dreaded dentist's drill and the equally dreaded practice of accountancy have all been shown to be part of Neolithic life is remarkable enough. Such early dates for these activities are not simply precocious oddities but part of a whole host of innovations and inventions of the Stone Age period. In the light of the vast body of evidence collected in this book, it is now clear that a fundamental reassessment of the prehistoric contribution to civilisation is necessary. Each of the elements of civilisation has been shown to have been highly developed long before the rise of ancient Egypt and Mesopotamia.

In the case of writing, we find that rather than being a unique innovation of Sumerian culture 5,000 years ago, it is an element of culture that grew out of prehistoric roots in various parts of the world. In the case of Sumeria and adjacent areas of the Near East, it has been shown that the cuneiform writing system was built on an earlier token system which has so far been traced back 10,000 years. Hieroglyphs used in the writings of Dynastic Egypt are now known to have been used in the prehistoric period on pottery and other artefacts, a thousand years before history began. Some scholars now predict that the origin of writing in Egypt may have to be put back in time, and this would mean that the date at which history itself begins would also have to be changed. As the study of prehistoric hieroglyphs in Egypt is pointing towards such a revolutionary change, so too the study of Stone Age signs on Chinese pottery may show that their system of writing is vastly older than is presently believed. There is also now a case for the independent existence of a written script of some kind in Old Europe, perhaps as early as 8,000 years ago.

Writing is the most fundamental element of civilisation, and its apparent absence in the Stone Age is seen as the best argument for maintaining the status quo. The evidence for a possibly much longer existence of the practice of writing in various parts of the world is becoming too large to simply ignore, and the apparently firm divide between history and prehistory is rapidly fading. For far from being over, the process of deciphering ancient signs has barely begun. A similar situation exists in the reconstruction of archaic languages. Until Sir William Jones demonstrated 200 years ago that Sanskrit was related to the European languages, such a thing would have been considered ridiculous. A number of historical linguists are now bringing forward evidence to show that the connections between language families are far deeper and earlier than had been previously thought. This indicates that such linguistic work, like that of decipherment, is far from over. Neither the Neolithic period nor even the Upper Palaeolithic era presents insurmountable barriers for the discovery of sign systems or etymologies which foreshadow later, historical developments.

It is not only the case that writing, still considered to be an innovation of the historical period, perhaps dates back to the Neolithic period. Other cultural elements once thought to be Neolithic in origin have now been traced back to Mesolithic and even Upper Palaeolithic times. Pottery, classically associated with Neolithic farmers, existed thousands of years earlier in the Jomon culture of Japan, and in Siberia about 13,000 years ago. Ceramic technology has been traced back *twice as far* by discoveries at the site of Dolni Vestonice in eastern Europe. Systematic underground mining, thought to have begun only comparatively late in the Neolithic era, has recently been shown to have existed 35,000 years ago in Egypt – almost at the beginning of the Upper Palaeolithic period. Similarly it was believed that the latter part of the Upper Palaeolithic period showed clear signs of having been more advanced than the early Upper Palaeolithic, for example in its works of art. The discovery of both cave art and figurines from the earliest phase of the Upper Palaeolithic shows an equal mastery of both drawing and sculpture to that of the later phases. Art and symbolism are still widely claimed by archaeologists to have originated in the Upper Palaeolithic period, that is to say no earlier than 40,000 years ago. Numerous examples of Middle Palaeolithic art and symbolic artefacts have been described in this book, and the recent verification of the Berekhat Ram figurine clearly shows that art began at least a quarter of a million years ago. The intentional engraving of bones has also been shown to have been undertaken as far back as the Lower Palaeolithic era, although once again, many archaeologists refuse to acknowledge the pre-Upper Palaeolithic evidence.

Preconceived opinions have repeatedly led to the rejection of evidence that does not fit with present archaeological dogmas. This has led to the routine acceptance of an Upper Palaeolithic engraved bone from the Hayonim Cave in Israel, and the equally routine rejection of an engraved bone of Middle Palaeolithic age from the same site, *even though the Middle Palaeolithic bone has more extensive markings than its Upper Palaeolithic counterpart*! A similar state of affairs existed in the case of the Tartaria tablets and related artefacts of Old Europe. When it was believed that they were later in date than Sumerian writing, it was seen as legitimate to treat them as possible writing systems, but when it became known that they preceded Sumerian civilisation, the whole question was dismissed purely on that basis. The same kind of prejudice has also occurred in the case of early sites in the Americas. The accepted dogma that there were no humans before about 12,000 years ago in either North or South America routinely led to the rejection of claims of earlier sites. There was a time when the idea that the cave paintings of Lascaux were of prehistoric age was thought ridiculous. In 1980 most archaeologists did not accept that there was evidence for human occupation of the Japanese islands before 30,000 years ago. Just over a decade later, archaeologists were already accepting that the initial occupation took place as early as 600,000 years ago.

This does not mean that all claims for early sites and artefacts should be automatically accepted without critical scrutiny. An impartial approach is what is required, and one unhampered by entrenched beliefs concerning what is and is not possible for any given era of prehistory. If a bone shows clear signs of having been engraved, its acceptance as a genuine artefact should be reached on the basis of the object itself and not merely on its age. There is overwhelming evidence that the whole conventional chronology for the various cultural innovations of mankind is fundamentally inaccurate. The basic sequence of stages in the chronological framework (Lower Palaeolithic, Middle Palaeolithic, Upper Palaeolithic, Mesolithic, Neolithic, Historical Civilisation) is not particularly problematic. The real problem lies in what cultural innovations are to be placed in each of these pigeonholes. I have already shown that there is good reason to question that the origin of writing should be placed in historical times, the origin of art in Upper Palaeolithic times, and so on. The balance needs to be tipped firmly backwards because many of the most fundamental cultural innovations actually occurred far earlier in the overall sequence than is generally realised.

The process of correcting the chronology of cultural events leads to the inevitable conclusion that the present division between history and prehistory is not as solid as it may appear. This new view of our prehistoric past is continually being strengthened as both new discoveries and new investigations of long-neglected artefacts are undertaken. In the process more and more evidence is tipping the scales further back and is showing that prehistoric cultural achievements are more profound, complex and multifarious than has been hitherto suspected. The full glory of prehistoric civilisation will never be known; too much has been lost forever due to the ravages of time and to human negligence and vandalism. Nevertheless, archaeological discoveries in the future will undoubtedly open many new doorways through which we may glimpse the reflected glory of the lost civilisations of the Stone Age.

Bibliography

Ackerknecht, E.H., 1978. 'Primitive Surgery', 164–80 in M.H. Logan and E.E. Hunt (eds.), *Health and the Human Condition: Perspectives on Medical Anthropology*, Duxbury Press, North Scituate, Massachusetts.

Adovasio, J.M., O. Soffer and B. Klíma, 1996. 'Upper Palaeolithic fibre technology: interlaced woven finds from Pavlov I, Czech Republic, *c.* 26,000 years ago', *Antiquity* 70, 526–34.

Alinei, M., 1981. 'More on Red Ochre: The Contribution of Diachronic Semantics', *Current Anthropology* 22/4, 443–4.

Allison, M.J. and E. Gerszten, 1982. *Paleopathology in South American Mummies: Applications of Modern Techniques*, Department of Pathology, Medical College of Virginia, Virginia Commonwealth University, Richmond, Virginia.

Alsozatai-Petheo, J., 1986. 'An Alternative Paradigm for the Study of Early Man in the New World', 15–26 in A.L. Bryan (ed.), *New Evidence for the Pleistocene Peopling of the Americas*, Center for the Study of Early Man, University of Maine, Orono, Maine.

Alt, K.W. *et al.*, 1997. 'Evidence for Stone Age Cranial Surgery', *Nature* 387, 360.

Althin, C.A., 1950. 'New Finds of Mesolithic Art in Scania (Sweden)', *Acta Archaeologica (Copenhagen)* 21, 253–60.

Anderson, A., 1994. 'Comment on J. Peter White's Paper "Site 820 and the Evidence for Early Occupation in Australia"', *Quaternary Australasia* 12/2, 30–1.

Anon., 1995. 'The Jomon Period Revisited', *The East* 31/6, 45–54.

Arnett, W.S., 1982. *The Predynastic Origin of Egyptian Hieroglyphics: Evidence for the Development of Rudimentary Forms of Hieroglyphs in Upper Egypt in the Fourth Millennium BC*, University Press of America, Washington DC.

Ashley-Montagu, M.F., 1937. 'The Origin of Subincision in Australia', *Oceania* 8/2, 193–207.

Auel, J.M., 1980. *The Clan of the Cave Bear*, Hodder and Stoughton, London.

Bahn, P.G., 1978. 'Water Mythology and the distribution of Palaeolithic Parietal Art', *Proceedings of the Prehistoric Society* 44, 125–34.

Bahn, P.G, 1986. 'No Sex, Please, We're Aurignacians', *Rock Art Research* 3/2, 99–120.

Bahn, P.G, 1987. 'Excavation of a palaeolithic plank from Japan', *Nature* 329, 110.

Bahn, P.G., 1991a. 'Pleistocene Images outside Europe', *Proceedings of the Prehistoric Society* 57/1, 91–102.

Bahn, P.G., 1991b. 'Dating the First American', *New Scientist* 20 July, 26–8.

Bahn, P.G., 1993. '50,000-year-old Americans of Pedra Furada', *Nature* 362, 114–15.

Bahn, P.G., 1996. 'New Developments in Pleistocene Art', *Evolutionary Anthropology* 4/6, 204–15.

Bahn, P.G. (ed), 1992. *Collins Dictionary of Archaeology*, Harper Collins, Glasgow.

Bahn, P.G. and J. Vertut, 1988. *Images of the Ice Age*, Facts On File, New York.

Bahn, P.G. and J. Vertut, 1997. *Journey through the Ice Age*, Weidenfeld and Nicolson, London.

Balter, M., 1996. 'Cave Structure Boosts Neanderthal Image', *Science* 271, 449.

Bar-Yosef, O. *et al.*, 'New Data on the Origin of Modern Man in the Levant', *Current Anthropology* 27/1, 63–4.

Bauval, R. and G. Hancock, 1996. *Keeper of Genesis: A Quest for the Hidden Legacy of Mankind*, Heinemann, London.

Beaune, S. A. de., 1987. 'Palaeolithic Lamps and Their Specialisation: A Hypothesis', *Current Anthropology* 28/4, 569–77.

Beaune, S. A. de., 1993. 'Nonflint Stone Tools of the Early Upper Paleolithic', 163–91 in H. Knecht, A. Pike-Tay and R. White (eds.), *Before Lascaux: The Complex Record of the Early Upper Paleolithic*, CRC Press, Ann Arbor.

Bednarik, R.G., 1989. 'The Galgenburg Figurine from Krems, Austria', *Rock Art Research* 6/2, 118–25.

Bednarik, R.G., 1990a. 'An Acheulian haematite pebble with striations', *Rock Art Research* 7/1, 75.

Bednarik, R.G., 1990b. 'More to Palaeolithic females than meets the eye', *Rock Art Research* 7/2, 133–7.

Bednarik, R.G., 1992. 'Palaeoart and Archaeological Myths', *Cambridge Archaeological Journal* 2/1, 27–57.

Bednarik, R.G., 1993. 'European Palaeolithic Art – Typical or Exceptional?', *Oxford Journal of Archaeology* 12/1, 1–8.

Bednarik, R.G., 1994a. 'The Pleistocene Art of Asia', *Journal of World Prehistory* 8/4, 351–75.

Bednarik, R.G., 1994b. 'A taphonomy of palaeoart', *Antiquity* 68, 68–74.

Bednarik, R.G., 1995. 'Concept-mediated Marking in the Lower Palaeolithic', *Current Anthropology* 36/4, 605–34.

Bednarik, R.G. and You Yuzhu, 1991. 'Palaeolithic Art from China', *Rock Art Research* 8/2, 119–23.

Belfer-Cohen, A. and O. Bar-Yosef, 1981. 'The Aurignacian at Hayonim Cave', *Paléorient* 7/2, 19–42.

Belfer-Cohen, A. and N. Goren-Inbar, 1994. 'Cognition and

Communication in the Levantine Lower Palaeolithic', *World Archaeology* 26/2, 144–57.

Belitzky, S., N. Goren-Inbar and E. Werker, 1991. 'A Middle Pleistocene wooden plank with man-made polish', *Journal of Human Evolution* 20, 349–53.

Bellwood, P., 1995. 'Language Families and Human Dispersal', *Cambridge Archaeological Journal* 5/2, 271–5.

Bengtson, J.D. and M. Ruhlen, 1994. 'Global Etymologies', 277–336 in M. Ruhlen (ed.), *On the Origins of Languages: Studies in Linguistic Taxonomy*, Stanford University Press, Stanford.

Bennike, P., 1985. *Palaeopathology of Danish Skeletons: A Comparative Study of Demography, Disease and Injury*, Akademisk Forlag, Copenhagen.

Blainey, G., 1983. *Triumph of the Nomads: A History of Ancient Australia*, Sun Books, Melbourne.

Blurton-Jones, N. and M.J. Konner, 1976. '!Kung Knowledge of Animal Behavior (or: The Proper Study of Mankind is Animals)', 325–48 in R.B. Lee and I. De Vore (eds.), *Kalahari Hunter-Gatherers: Studies of the !Kung San and Their Neighbors*, Harvard University Press, Cambridge, Massachusetts.

Bock, W.J., A. Boldt-Gäth and R. Meschig, 1986. 'From the Beginning of Trepanation to Modern Neurochirurgie (from the exhibition catalogue of the XXX International Congress of the History of Medicine, Düsseldorf, 6 August to 7 September 1986).

Borrero, L.A., 1995. 'Human and natural agency: some comments on Pedra Furada', *Antiquity* 69, 602–3.

Bouissac, P., 1994a. 'Introduction: A challenge for semiotics', *Semiotica* 100/2–4 Special Issue: Prehistoric Signs, 99–107.

Bouissac, P., 1994b. 'Art or script? A falsifiable semiotic hypothesis', *Semiotica* 100/2–4 Special Issue: Prehistoric Signs, 349–67.

Boyd, B. and J. Cook, 1993. 'A Reconsideration of the "Ain Sakhri" Figurine', *Proceedings of the Prehistoric Society*, 59, 399–405.

Brodie, F., 1971. *The Devil Drives: A Life of Sir Richard Burton*, Penguin, Harmondsworth.

Brothwell, D.R., 1978. 'The question of pollution in earlier and less developed societies', 129–36 in M.H. Logan and E.E. Hunt (eds.), *Health and the Human Condition*, Duxbury Press, North Scituate, Massachusetts.

Brothwell, D.R., 1991. 'On zoonoses and their relevance to palaeopathology', 18–22 in D.J. Ortner and A.C. Aufderheide, *Human Palaeopathology: Current Syntheses and Future Options*, Smithsonian Institution Press, Washington DC.

Brothwell, D.R., 1994. 'Ancient Trephining: Multi-focal Evolution or Trans-World Diffusion?', *Journal of Paleopathology* 6/3, 129–38.

Bryan, A.L., 1986. 'Paleoamerican Prehistory as Seen from South

America', 1–14 in A.L. Bryan (ed.), *New Evidence for the Pleistocene Peopling of the Americas*, Center for the Study of Early Man, University of Maine, Orono, Maine.

Burl, A., 1983. *Prehistoric Astronomy and Ritual*, Shire Publications, Princes Risborough, Aylesbury.

Burton, R.F., 1864. 'Notes on Scalping', *Anthropological Review* 2, 49–52.

Bynon, T., 1995. 'Can there Ever be a Prehistorical Linguistics?', *Cambridge Archaeological Journal* 5/2, 261–5.

Carter, G.F., 1980. 'The Metate: An Early Grain-Grinding Implement in the New World', 21–39 in D.L. Browman (ed.), *Early Native Americans: Prehistoric Demography, Economy, and Technology*, Mouton, The Hague.

Cave-Browne, P. (n.d.). *Fire-Making: A Survival Skill from the Past*, Pitt Rivers Museum, Oxford.

Chase, P.G., 1991. 'Symbols and Paleolithic Artifacts: Style, Standardisation, and the Imposition of Arbitrary Form', *Journal of Anthropological Archaeology* 10, 193–214.

Chase, P.G. and H.L. Dibble, 1987. 'Middle Paleolithic Symbolism: A Review of Current Evidence and Interpretations', *Journal of Anthropological Archaeology* 6, 263–6.

Cheetham, L., 1987. *Making Light Work*, Pitt Rivers Museum, Oxford.

Chilardi, S. *et al.*, 1996. 'Fontana Nuova di Ragusa (Sicily, Italy): southernmost Aurignacian site in Europe', *Antiquity* 70, 553–63.

Clarke, R., 1935. 'The Flint-Knapping Industry at Brandon', *Antiquity* 9, 38–56.

Clottes, J., 1996. 'Thematic changes in Upper Palaeolithic art: a view from the Grotte Chauvet', *Antiquity* 70, 276–88.

Clottes, J., M. Menu and P. Walter, 1990. 'New Light on the Niaux Paintings', *Rock Art Research* 7/1, 21–6.

Clutton-Brock, J., 1984. *Neolithic Antler Picks from Grimes Graves, Norfolk, and Durrington Walls, Wiltshire: A Biometrical Analysis*, Excavations at Grimes Graves, Norfolk 1972–1976 (Fascicule 1), British Museum Publications, London.

Cohen, M.N. and G.J. Armelagos (eds.), 1984. *Paleopathology at the Origins of Agriculture*, Academic Press, New York.

Cole, J.R., 1980. 'Cult Archaeology and Unscientific Method and Theory', *Advances in Archaeological Method and Theory* 3, 1–33.

Conkey, M., 1983. 'On the Origins of Paleolithic Art: A Review and Some Critical Thoughts', 201–27 in E. Trinkaus (ed.), *The Mousterian Legacy: Human Biocultural Change in the Upper Pleistocene*, British Archaeological Reports, International Series 164, Oxford.

Cotterell, B. and J. Kamminga, 1990. *Mechanics of Pre-Industrial Technology*, Cambridge University Press, Cambridge.

Courville, C.B., 1967. 'Cranial Injuries in Prehistoric Man', 606–22 in D. Brothwell and A.T. Sandison (eds.), *Diseases in Antiquity: A Survey of the Diseases, Injuries and Surgery of Early Populations*, Charles C. Thomas, Springfield, Illinois.

Cowan, H.K.J., 1975. 'More on Upper Paleolithic Engraving', *Current Anthropology* 16/2, 297–8.

Cranstone, B.A.L., 1951. 'Stone Age Man's Use of Power', *Man* (April Issue), 48–50.

Cremo, M.A. and R.L. Thompson, 1993. *Forbidden Archeology: The Hidden History of the Human Race*, Bhaktivedanta Institute, San Diego.

Dams, L., 1984. 'Preliminary Findings at the "Organ Sanctuary" in the Cave of Nerja, Malaga, Spain', *Oxford Journal of Archaeology* 3/1, 1–14.

Dams, L., 1985. 'Palaeolithic Lithophones: Descriptions and Comparisons', *Oxford Journal of Archaeology* 4/1, 31–46.

Davidson, D.S., 1947. 'Fire-Making in Australia', *American Anthropologist* 49/3, 426–37.

Davidson, I., 1990. 'Bilzingsleben and early marking', *Rock Art Research* 7/1, 52–6.

Davis, S., 1974. 'Incised Bones from the Mousterian of Kebara Cave (Mount Carmel) and the Aurignacian of Ha-Yonim Cave (Western Galilee), Israel', *Paléorient* 2/1, 181–2.

Delporte, H., 1993. 'Gravettian Female Figurines: A Regional Survey', 243–57 in II. Knecht, A. Pike-Tay and R. White (eds.), *Before Lascaux: The Complex Record of the Early Upper Paleolithic*, CRC Press, Ann Arbor.

Dennell, R.W., 1989. 'Reply', *Current Anthropology* 30/3, 318–22.

Dennell, R.W. and L. Hurcombe, 1995. 'Comment on Pedra Furada', *Antiquity* 69, 604.

Dennell, R.W., H.M. Rendell and E. Hailwood, 1988. 'Late Pliocene Artefacts from Northern Pakistan', *Current Anthropology* 29/3, 495–8.

d'Errico, F., 1989. 'Palaeolithic Lunar Calendars: A Case of Wishful Thinking?', *Current Anthropology* 30/1, 117–18.

d'Errico, F., 1991. 'Microscopic and Statistical Criteria for the Identification of Prehistoric Systems of Notation', *Rock Art Research* 8/2, 83–93.

d'Errico, F., 1992. 'A Reply to Alexander Marshack', *Rock Art Research* 9/1, 59–64.

d'Errico, F., 1995. 'A New Model and its Implications for the Origin of Writing: The La Marche Antler Revisited', *Cambridge Archaeological Journal* 5/2, 163–206.

d'Errico, F. and C. Cacho, 1994. 'Notation Versus Decoration in the Upper Palaeolithic: a Case-Study from Tossal de la Roca, Alicante, Spain', *Journal of Archaeological Science* 21, 185–200.

Dewez, M.C., 1974. 'New hypotheses concerning two engraved bones from La Grotte de Remouchamps, Belgium', *World Archaeology* 5/3, 337–45.

Diringer, D., 1962. *Writing*, Thames and Hudson, London.

Dolgopolsky, A., 1995. 'Linguistic Prehistory', *Cambridge Archaeological Journal* 5/2, 268–71.

Duff, A.I., G.A. Clark and T.J. Chadderdon, 1992. 'Symbolism in the Early Palaeolithic: A Conceptual Odyssey', *Cambridge Archaeological Journal* 2/2, 211–29.

Eaton, S.B., M. Shostak and M. Konner, 1989. *The Stone-Age Health Programme: Diet and Exercise as Nature Intended*, Angus and Robertson, London.

Edwards, S.W., 1978. 'Nonutilitarian Activities in the Lower Paleolithic: A Look at the Two Kinds Of Evidence', *Current Anthropology* 19/1, 135–7.

Eliade, M., 1972. *Shamanism: Archaic Techniques of Ecstasy*, Princeton University Press, Princeton.

Eliade, M., 1978. *The Forge and the Crucible: The Origins and Structures of Alchemy*, University of Chicago Press, Chicago.

Epstein, J.F., 1979. 'Flint Technology and the Heating of Stone', 27–38 in D. Schmandt-Besserat (ed.), *Early Technologies*, Undeno Publications, Malibu.

Evans, J.D., 1959. *Malta*, Thames and Hudson, London.

Fairservis, W.A., 1983. *Hierakonpolis: The Graffiti and the Origins of Egyptian Hieroglyphic Writing*, The Hierakonpolis Project, Occasional Papers in Anthropology Number II, Vassar College, Poughkeepsie, NY.

Farmer, M.F., 1994. 'The Origins of Weapon Systems', *Current Anthropology* 35/5, 679–81.

Felkin, R.W., 1884. 'Notes on Labour in Central Africa', *Edinburgh Medical Journal* 29, 922–30.

Fiennes, R.N.T-W., 1978. *Zoonoses and the Origins and Ecology of Human Disease*, Academic Press, London.

Fischer, A., 1974. 'An Ornamented Flint-Core from Holmegård V, Zealand, Denmark: Notes on Mesolithic Ornamentation and Flint-Knapping', *Acta Archaeologica (Copenhagen)* 45, 155–68.

Flood, J., 1983. *Archaeology of the Dreamtime*, Collins, London.

Forbes, A. and T.R. Crowder, 1979. 'The problem of Franco-Cantabrian abstract signs: agenda for a new approach', *World Archaeology* 10/3, 350–66.

Frolov, B.A., 1978a. 'Numbers in Paleolithic Graphic Art and the Initial Stages in the Development of Mathematics, Part One', *Soviet Anthropology and Archeology* 16/3–4, 142–66.

Frolov, B.A., 1978b. 'Numbers in Paleolithic Graphic Art and the Initial Stages in the Development of Mathematics, Part Two', *Soviet Anthropology and Archeology* 17/1, 73–93.

Frolov, B.A., 1979a. 'Numbers in Paleolithic Graphic Art and the Initial Stages in the Development of Mathematics, Part Three', *Soviet Anthropology and Archeology* 17/3, 41–74.

Frolov, B.A., 1979b. 'Numbers in Paleolithic Graphic Art and the Initial Stages in the Development of Mathematics, Part Four', *Soviet Anthropology and Archeology* 17/4, 61–113.

Frolov, B.A., 1981. 'On Astronomy in the Stone Age', *Current Anthropology* 22/5, 585.

Fullagar, R. and J. Field, 1997. 'Pleistocene seed-grinding implements from the Australian arid zone', *Antiquity* 71, 300–7.

Fullagar, R.L.K., D.M. Price and L.M. Head, 1996. 'Early human occupation of northern Australia: archaeology and thermoluminescence dating of Jinmium rock-shelter, Northern Territory', *Antiquity* 70, 751–73.

Gargett, R.H., 1989. 'Grave Shortcomings: The Evidence for Neanderthal Burial', *Current Anthropology* 30/2, 157–90 (and discussions in 30/3, 322–30 by various authors).

Gimbutas, M., 1982. *The Goddesses and Gods of Old Europe: 6500–3500 BC, Myths and Cult Images*, Thames and Hudson, London.

Gimbutas, M., 1989. *The Language of the Goddess: Unearthing the Hidden Symbols of Western Civilisation*, Thames and Hudson, London.

Gimbutas, M., 1991. *The Civilisation of the Goddess: The World of Old Europe*, edited by J. Marler, Thames and Hudson, London.

Ginzburg, C., 1990. 'Clues: Roots of an Evidential Paradigm', 96–125 in *Myths, Emblems, Clues*, trans. J. and A.C. Tedeschi, Hutchinson Radius, London.

Goren-Inbar, N., 1985. 'The Lithic Assemblage of the Berekhat Ram Acheulian Site, Golan Heights', *Paléorient* 11/1, 7–28.

Goren-Inbar, N., 1986. 'A Figurine from the Acheulian Site of Berekhat Ram', *Mitekufat Haeven (Journal of the Israel Prehistoric Society)* NS 19, 7–12.

Goren-Inbar, N. and S. Peltz, 1995. 'Additional remarks on the Berekhat Ram figurine', *Rock Art Research* 12/2, 131–2.

Gould, R.A., 1977. 'A Case of Heat Treatment of Lithic Materials in Aboriginal Northwestern California', *Journal of California Anthropology* 4/1, 142–4.

Gowlett, J.A.J., 1984. *Ascent to Civilisation: The Archaeology of Early Man*, William Collins, London.

Grimal, N., 1992. *A History of Ancient Egypt*, trans. I. Shaw, Blackwell, Oxford.

Gruhn, R., 1987. 'On the Settlement of the Americas: South American Evidence for an Expanded Time Frame', *Current Anthropology* 28/3, 363–4.

Guidon, N. and B. Arnaud, 1991. 'The chronology of the New World: two faces of one reality', *World Archaeology* 23/2, 167–78.

Guidon, N. and G. Delibrias, 1986. 'Carbon-14 dates point to man in the Americas 32,000 years ago', *Nature* 321, 769–71.

Guidon, N. *et al.*, 1996. 'Nature and age of the deposits in Pedra Furada, Brazil: reply to Meltzer, Adovasio & Dillehay', *Antiquity* 70, 408–21.

Haarmann, H., 1989. 'Writing From Old Europe to Ancient Crete – A Case of Cultural Continuity', *The Journal of Indo-European Studies*, 17/3–4, 251–75.

Habgood, P.J., 1989. 'Bilzingsleben: to be or not to be *Homo erectus*', *Rock Art Research* 6/2, 139–41.

Hahn, J., 1993. 'Aurignacian Art in Central Europe', 229–42 in H. Knecht, A. Pike-Tay and R. White (eds.), *Before Lascaux: The Complex Record of the Early Upper Paleolithic*, CRC Press, Ann Arbor.

Hamilton, A., 1980. 'Dual Social Systems: Technology, Labour and Women's Secret Rites in the eastern Western Desert of Australia', *Oceania* 51/1, 4–19.

Hamperl, H., 1967. 'The Osteological Consequences of Scalping', 630–4 in D. Brothwell and A.T. Sandison (eds.), *Diseases in Antiquity: Diseases, Injuries and Surgery in Early Populations*, Charles C. Thomas, Springfield, Illinois.

Harner, M.J., 1990. *The Way of the Shaman*, Harper and Row, New York.

Harrison, H.S., 1954. 'Fire-Making, Fuel, and Lighting', 216–37 in C. Singer, E.J. Holmyard and A.R. Hall (eds.), *A History of Technology, Volume I: From Early Times to Fall of Ancient Empires*, Clarendon Press, Oxford.

Harrold, F.B., 1980. 'A comparative analysis of Eurasian Palaeolithic burials', *World Archaeology* 12/2, 195–211.

Hassan, F.A., 1983. 'The Roots of Egyptian Writing', *Quarterly Review of Archaeology* 4/3, 1, 7–8.

Hassan, F.A., 1984. 'The Beginnings of Egyptian Civilisation at Hierakonpolis', *Quarterly Review of Archaeology* 5/1, 13–15.

Hassan, F.A., 1988. 'The Predynastic of Egypt', *Journal of World Prehistory* 2/2, 135–85.

Heggie, D.C., 1981. *Megalithic Science: Ancient Mathematics and Astronomy in North-West Europe*, Thames and Hudson, London.

Heggie, D.C., 1982. 'Megalithic Astronomy: Highlights and Problems', 1–24 in D.C. Heggie (ed.), *Archaeoastronomy in the Old World*, Cambridge University Press, Cambridge.

Hemingway, M.F., 1989. 'Early Artefacts from Pakistan? Some Questions for the Excavators', *Current Anthropology* 30/3, 317–18.

Henriksen, G., 1973. 'Maglemosekulturens Drilbor med et par boretekniske betragtninger (Drilling in the Maglemose Culture)', *Aarbøger for nordisk Oldkyndighed og Historie 1973*, 217–25.

Holgate, R., 1991. *Prehistoric Flint Mines*, Shire Publications, Princes Risborough, Aylesbury.

Hood, M.S.F., 1967. 'The Tartaria Tablets', *Antiquity* 41, 99–113.

Hovers, E., Y. Rak and W.H. Kimbel, 1996. 'Neanderthals of the Levant: A baby's burial sheds light on the development and behavior of the species', *Archaeology* Jan–Feb. issue, 49–50.

Huyge, D., 1990. 'Mousterian Skiffle? Note on a Middle Palaeolithic Engraved Bone from Schulen, Belgium', *Rock Art Research* 7/2, 125–32.

Huyge, D., 1991 'The "Venus" of Laussel in the Light of Ethnomusicology', *Archeologie in Vlaanderen* 1, 11–18.

Imamura, K., 1996. *Prehistoric Japan: New perspectives on insular East Asia*, UCL Press, London.

James, S.R., 1989. 'Hominid Use of Fire in the Lower and Middle Pleistocene: A Review of the Evidence', *Current Anthropology* 30/1, 1–26.

James, T.G.H., 1979. *An Introduction to Ancient Egypt*, British Museum Publications, London.

Janssens, P., 1970. *Palaeopathology: Diseases and Injuries of Prehistoric Man*, John Baker, London.

Jia, Lanpo, 1985. 'China's Earliest Palaeolithic Assemblages', 135–45 in Wu Rukang and J.W. Olsen (eds.), *Palaeoanthropology and Palaeolithic Archaeology in the People's Republic of China*, Academic Press, London.

Johanson, D.C. and M.A. Edey, 1981. *Lucy: The Beginnings of Humankind*, Simon and Schuster, New York.

Johnson, L.L., 1978. 'A History of Flint-Knapping Experimentation, 1838–1976', *Current Anthropology* 19/2, 337–72.

Keeley, L.H., 1980. *Experimental Determination of Stone Tool Uses: A Microwear Analysis*, University of Chicago Press, Chicago.

Kennedy, D.G., 1929. 'Field Notes on the Culture of Vaitupu, Ellice Islands', Memoir Supplement in *Journal of the Polynesian Society* 38, 1–38.

Kenrick, D.M., 1995. *Jomon of Japan: The World's Oldest Pottery*, Kegan Paul International, London.

Kershaw, A.P., 1994. 'Site 820 and the Evidence for Early Occupation in Australia – A Response', *Quaternary Australasia* 12/2, 24–9.

Klein, R.G., 1995. 'Anatomy, Behavior, and Modern Human Origins', *Journal of World Prehistory* 9/2, 167–98.

Knecht, H., A. Pike-Tay and R. White, 1993. 'Introduction', 1–4 in H. Knecht, A. Pike-Tay and R. White (eds.), *Before Lascaux: The Complex Record of the Early Upper Paleolithic*, CRC Press, Ann Arbor.

Kraybill, N., 1977. 'Pre-Agricultural Tools for the Preparation of Foods in the Old World', 485–521 in C.A. Reed (ed.), *Origins of Agriculture*, Mouton, The Hague.

Kroeber, A.L., 1948. *Anthropology: Race, Language, Culture, Psychology, Prehistory*, Harcourt, Brace and Company, New York.

Kumar, G., 1996. 'Daraki-Chattan: A Palaeolithic Cupule Site in India', *Rock Art Research* 13/1, 38–46.

Larichev, V., U. Khol'ushkin and I. Laricheva, 1987. 'Lower and Middle Paleolithic of Northern Asia: Achievements, Problems, and Perspectives', *Journal of World Prehistory* 1/4, 415–64.

Larichev, V., U. Khol'ushkin and I. Laricheva, 1988. 'The Upper Paleolithic of Northern Asia: Achievements, Problems, and Perspectives. I. Western Siberia', *Journal of World Prehistory* 2/4, 359–96.

Larichev, V., U. Khol'ushkin and I. Laricheva, 1990. 'The Upper Paleolithic of Northern Asia: Achievements, Problems, and Perspectives. II. Central and Eastern Siberia', *Journal of World Prehistory* 4/3, 347–85.

Larsson, L., 1990. 'The Mesolithic of Southern Scandinavia', *Journal of World Prehistory* 4/3, 257–309.

Leakey, M.D., 1971. *Olduvai Gorge, Volume 3: Excavations in Beds I and II, 1960–63*, Cambridge University Press, Cambridge.

Leroi-Gourhan, A., 1982. *The Dawn of European Art: An Introduction to Palaeolithic Cave Painting*, Cambridge University Press, Cambridge.

Leroi-Gourhan, A., 1975. 'The Flowers Found with Shanidar IV, a Neanderthal Burial in Iraq', *Science* 190, 562–4.

Liebenberg, L., 1990. *The Art of Tracking: The Origin of Science*, David Philip, Claremont, South Africa.

Lietava, J., 1992. 'Medicinal plants in a Middle Paleolithic grave Shanidar IV?', *Journal of Ethnopharmacology* 35/3, 263–6.

Lisowski, F.P., 1967. 'Prehistoric and Early Historic Trepanation', 651–72 in D. Brothwell and A.T. Sandison (eds.), *Diseases in Antiquity: A Survey of the Diseases, Injuries and Surgery of Early Populations*, Charles C. Thomas, Springfield, Illinois.

Loeb, E.M., 1926. 'Pomo Folkways', *University of California Publications in Archaeology and Ethnology* 19/2, 149–404.

Longworth, I., *et al.*, 1991. *Excavations at Grimes Graves, Norfolk 1972–1976 Fascicule 3; Shaft X: Bronze Age Flint, Chalk and Metal Working*, British Museum Publications, London.

Lorblanchet, M., *et al.*, 1990. 'Palaeolithic Pigments in the Quercy, France', *Rock Art Research* 7/1, 4–20.

McDermott, L., 1996. 'Self-Representation in the Upper Paleolithic Female Figurines', *Current Anthropology* 37/2, 227–75.

McKeown, T., 1988. *The Origins of Human Disease*, Blackwell, Oxford.

MacRae, R.J., 1988. 'Belt, Shoulder-Bag or Basket? An Enquiry into Handaxe Transport and Flint Sources', *Lithics* 9, 2–8.

Makkay, J., 1968. 'The Tartaria Tablets', *Orientalia* NS 37, 272–89.

Mania, D. and U. Mania, 1988. 'Deliberate Engravings on Bone Artefacts of Homo Erectus', *Rock Art Research* 5/2, 91–107.

Mania, D. and U. Mania, 1989. 'Reply to Habgood', *Rock Art Research* 6/2, 141–2.

Margetts, E.L., 1967. 'Trepanation of the Skull by the Medicine-men of Primitive Cultures, with Particular Reference to Present-day Native East African Practice', 673–701 in D. Brothwell and A.T. Sandison (eds.), *Diseases in Antiquity: A Survey of the Diseases, Injuries and Surgery of Early Populations*, Charles C. Thomas, Springfield, Illinois.

Marsh, G.H. and W.S. Laughlin, 1956. 'Human Anatomical Knowledge among the Aleutian Islanders', *Southwestern Journal of Anthropology* 12/1, 38–78.

Marshack, A., 1972. *The Roots of Civilisation*, McGraw Hill, New York (2nd ed., 1991).

Marshack, A., 1975. 'Reply to Cowan', *Current Anthropology* 16/2, 298.

Marshack, A., 1976. 'Some Implications of the Paleolithic Symbolic Evidence for the Origin of Language', *Current Anthropology* 17/2, 274–82.

Marshack, A., 1979. 'Upper Paleolithic Symbol Systems of the Russian Plain: Cognitive and Comparative Analysis', *Current Anthropology* 20/2, 271–311.

Marshack, A., 1981. 'On Paleolithic Ochre and the Early Uses of Color and Symbol', *Current Anthropology* 22/2, 188–91.

Marshack, A., 1985. 'Theoretical concepts that lead to new analytical methods, modes of inquiry and classes of data', *Rock Art Research* 2/2, 95–111.

Marshack, A., 1991a. 'The Female Image: A "Time-factored" Symbol. A Study in Style and Aspects of Image Use in the Upper Palaeolithic', *Proceedings of the Prehistoric Society* 57/1, 17–31.

Marshack, A., 1991b. 'A reply to Davidson on Mania and Mania', *Rock Art Research* 8/1, 47–58.

Marshack, A., 1991c. 'The Taï Plaque and Calendrical Notation in the Upper Palaeolithic', *Cambridge Archaeological Journal* 1/1, 25–61.

Marshack, A., 1992. 'An innovative analytical technology: discussion of its present and potential use', *Rock Art Research* 9/1, 37–59.

Marshack, A., 1996. 'A Middle Paleolithic Symbolic Composition from the Golan Heights: The Earliest Known Depictive Image', *Current Anthropology* 37/2, 357–65.

Marshack, A., 1997. 'The Berekhat Ram figurine: a late Acheulian carving from the Middle East', *Antiquity* 71, 327–37.

Megarry, T., 1995. *Society in Prehistory: The Origins of Human Culture*, Macmillan, London.

Megaw, J.V.S., 1968. 'Problems and non-problems in palaeo-organology: a musical miscellany', 333–58 in J.M. Coles and D.D.A. Simpson (eds.), *Studies in Ancient Europe: Essays presented to Stuart Piggott*, Leicester University Press, Leicester.

Mellaart, J., 1967. *Çatal Hüyük: A Neolithic Town in Anatolia*, Thames and Hudson, London.

Mellars, P., 1989. 'Major Issues in the Emergence of Modern Humans', *Current Anthropology* 30/3, 349–85.

Meltzer, D.J., J.M. Adovasio and T.D. Dillehay, 1994. 'On a Pleistocene human occupation at Pedra Furada, Brazil', *Antiquity* 68, 695–714.

Mercer, R.J., 1981. 'Summary of the Excavation', vi–ix in *Grimes Graves, Norfolk: Excavations 1971–72*, (Volume II), Department of the Environment Archaeological Reports No. 11, HMSO, London.

Mészáros, G. and L. Vértes, 1955. 'A Paint Mine from the Early Upper Palaeolithic Age near Lovas (Hungary, County Veszprém)', *Acta Archaeologica Academiae Scientiarum Hungaricae* 5, 1–32.

Mishra, S. *et al.*, 1995. 'Earliest Acheulian Industry from Peninsular India', *Current Anthropology* 36/5, 847–51.

Mithen, S., 1996. 'On Early Palaeolithic "Concept-mediated Marks",
  Mental Modularity, and the Origins of Art', *Current Anthropology* 37/4,
  666–70.

Mortlock, J., 1997. 'First hearth, first picture', *East Anglian Daily Times* 11
  August, 9.

Movius, H.L., 1953. 'The Mousterian cave of Teshik-Tash, Southeastern
  Uzbekistan, Central Asia', *American School of Prehistoric Research* 17,
  11–71.

Needham, J., 1954. *Science and Civilisation in China, Volume 1:
  Introductory Orientations*, Cambridge University Press, Cambridge.

Needham, J., 1965. *Science and Civilisation in China, Volume 4: Physics
  and Physical Technology (Part II: Mechanical Engineering)*, Cambridge
  University Press, Cambridge.

Neustupný, E., 1968. 'The Tartaria Tablets: A Chronological Issue',
  Antiquity 42, 32–5.

Nissen, H.J., P. Damerow and R.K. Englund, 1993. *Archaic Bookkeeping:
  Early Writing and Techniques of Economic Administration in the Ancient
  Near East*, trans. P. Larsen, University of Chicago Press, Chicago and
  London.

Oakley, K.P., 1973. 'Fossil shell observed by Acheulian man', *Antiquity* 47,
  59–60.

Okladnikov, A.P. and G.A. Pospelova, 1982. 'Ulalinka, the Oldest
  Palaeolithic Site in Siberia', *Current Anthropology* 23/6, 710–12.

Parry, T.W., 1914. 'Prehistoric Man and His Early Efforts to Combat
  Disease', *The Lancet* 13 June.

Parry, T.W., 1916. 'The Art of Trephining Among Prehistoric and Primitive
  Peoples: Their Motives for its Practice and Their Methods of Procedure',
  Journal of the British Archaeological Association March 1916, 33–69.

Parry, T.W., 1918. *Surgery of the Stone Age: A Ballad of Neolithic Times*,
  John Bale, London.

Parry, T.W., 1923. *Trephination of the Living Human Skull in Prehistoric
  Times*, British Medical Association, London.

Pasquale, A. de., 1984. 'Pharmacognosy: The Oldest Modern Science',
  Journal of Ethnopharmacology 11, 1–16.

Paterson, L.W., 1983. 'Criteria for Determining the Attributes of Man-Made
  Lithics', *Journal of Field Archaeology* 10, 297–307.

Pelcin, A., 1994. 'A Geological Explanation for the Berekhat Ram
  Figurine', *Current Anthropology* 35/5, 674–5.

Petrie, W.M.F., 1912. *The Formation of the Alphabet*, Macmillan, London.

Piggott, S., 1940. 'A Trepanned Skull of the Beaker Period from Dorset and
  the Practice of Trepanning in Prehistoric Europe', *Proceedings of the
  Prehistoric Society* 6, 112–32.

Pitts, M. and M. Roberts, 1997. *Fairweather Eden: Life in Britain half a
  million years ago as revealed by the excavations at Boxgrove*, Century,
  London.

Purdy, B.A., 1982. 'Pyrotechnology: Prehistoric Application to Chert Materials in North America', 31–44 in T.A. and S.F. Wertime (eds.), *Early Pyrotechnology: The Evolution of the First Fire-Using Industries*, Smithsonian Institution Press, Washington DC.

Renfrew, C., 1976. *Before Civilisation: The Radiocarbon Revolution and Prehistoric Europe*, Penguin, Harmondsworth.

Renfrew, C., 1991. 'Before Babel: Speculations on the Origins of Linguistic Diversity', *Cambridge Archaeological Journal* 1/1, 3–23.

Renfrew, C., 1992. 'Archaeology, Genetics and Linguistic Diversity', *Man* 27/3, 445–78.

Renfrew, C., 1995. 'Towards a New Synthesis?', *Cambridge Archaeological Journal* 5/2, 258–61.

Rigaud, J-P, J.F. Simek and T. Ge, 1995. 'Mousterian fires from Grotte XVI (Dordogne, France)', *Antiquity* 69, 902–12.

Rivière, P., 1989. 'Body-snatching in South America', *Anthropology Today* 5/6, 23.

Rozoy, J-G., 1990. 'The Revolution of the Bowmen in Europe', 13–28 in C. Bonsall (ed.), *The Mesolithic in Europe: Papers Presented at the Third International Symposium, Edinburgh 1985*, John Donald, Edinburgh.

Ruggles, C.L.N., 1989. 'Recent developments in megalithic astronomy', 13–26 in A.F. Aveni (ed.), *World archaeoastronomy: Selected papers from the 2nd Oxford International Conference on Archaeoastronomy Held at Merida, Yucatan, Mexico 13–17 January 1986*, Cambridge University Press, Cambridge.

Ruhlen, M., 1987. *A Guide to the World's Languages, Volume One: Classification*, Edward Arnold, London.

Ruhlen, M., 1995. 'Linguistic Evidence for Human Prehistory', *Cambridge Archaeological Journal* 5/2, 265–8.

Schick, K.D. and N. Toth, 1993. *Making Silent Stones Speak: Human Evolution and the Dawn of Technology*, Touchstone, New York.

Schmandt-Besserat, D., 1980. 'Ocher in Prehistory: 300,000 Years of the Use of Iron Ore as Pigments', 127–50 in T.A. Wertime and J.D. Muhly (eds.), *The Coming of the Age of Iron*, Yale University Press, New Haven.

Schmandt-Besserat, D., 1992. *Before Writing, Volume One: From Counting to Cuneiform*, University of Texas Press, Austin.

Schmid, E., 1972. 'A Mousterian silex mine and dwelling-place in the Swiss Jura', 129–32 in F. Bordes (ed.), *The Origin of Homo Sapiens* (Ecology and Conservation 3), Unesco, Paris.

Schuiling, W.C. (ed.), 1972. *Pleistocene Man at Calico: A Report on the International Conference on the Calico Mountains Excavations, San Bernardino County, California*, San Bernardino County Museum Association, San Bernardino.

Seidenberg, A., 1962. 'The Ritual Origin of Counting', *Archive for History of Exact Sciences* 2/1, 1–40.

Shepherd, R., 1980. *Prehistoric Mining and Allied Industries*, Academic Press, London.

Shevoroshkin, V.V. and T.L. Markey (eds.), 1986. *Typology, Relationship and Time: A Collection of Papers on Language Change and Relationship By Soviet Linguists*, Karoma, Ann Arbor.

Simpson, R.D., 1980. 'Calico Mountains Site: Pleistocene Archaeology in the Mojave Desert, California', 7–20 in D.L. Browman (ed.), *Early Native Americans: Prehistoric Demography, Economy, and Technology*, Mouton, The Hague.

Simpson, R.D., L.W. Patterson and C.A. Singer, 1986. 'Lithic Technology of the Calico Mountains Site, Southern California', 89–105 in A.L. Bryan (ed.), *New Evidence for the Pleistocene Peopling of the Americas*, Center for the Study of Early Man, University of Maine, Orono, Maine.

Skertchly, S.B.J., 1879. *On the Manufacture of Gun-Flints, The Methods of Excavating for Flint, The Age of Palaeolithic Man, and the Connexion Between Neolithic Art and the Gun-Flint Trade*, Memoirs of the Geological Survey, HMSO, London.

Smirnov, Y., 1989. 'Intentional Human Burial: Middle Palaeolithic (Last Glaciation) Beginnings', *Journal of World Prehistory* 3/2, 199–233.

Smolla, G., 1987. 'Prehistoric flint mining: the history of research – a review', 127–9 in G. de G. Sieveking and M.H. Newcomer (eds.), *The Human Uses of Flint and Chert*, Cambridge University Press, Cambridge.

Soffer, O., *et al.*, 1993. 'The Pyrotechnology of Performance Art: Moravian Venuses and Wolverines', 259–75 in H. Knecht, A. Pike-Tay and R. White (eds.), *Before Lascaux: The Complex Record of the Early Upper Paleolithic*, CRC Press, Ann Arbor.

Solecki, R.S., 1972. *Shanidar: The Humanity of Neanderthal Man*, Allen Lane (Penguin), London.

Solecki, R.S., 1975. 'Shanidar IV, a Neanderthal Flower Burial in Northern Iraq', *Science* 190, 880–1.

Solecki, R.S., 1977. 'The Implications of the Shanidar Cave Neanderthal Flower Burial', *Annals of the New York Academy of Sciences* 293, 114–24.

Srejovic, D., 1972. *Europe's First Monumental Sculpture: New Discoveries at Lepenski Vir*, Thames and Hudson, London.

Stapert, D., 1989. 'Early Artefacts from Pakistan? Some Questions for the Excavators', *Current Anthropology* 30/3, 318.

Steensberg, A., 1980. *New Guinea Gardens: A Study of Husbandry with Parallels in Prehistoric Europe*, Academic Press, London.

Steensberg, A., 1986. *Man the Manipulator: An Ethno-Archaeological Basis for Reconstructing the Past*, National Museum of Denmark, Copenhagen.

Stepanchuk, V.N., 1993. 'Prolom II, a Middle Palaeolithic Cave Site in the Eastern Crimea with Non-Utilitarian Bone Artefacts', *Proceedings of the Prehistoric Society* 59, 17–37.

Taçon, P.S.C. *et al.*, 1997. 'Cupule engravings from Jinmium-Granilpi (northern Australia) and beyond: exploration of a widespread and enigmatic class of rock markings', *Antiquity* 71, 942–65.

Taylor, T., 1996. *The Prehistory of Sex: Four Million Years of Human Sexual Culture*, Fourth Estate, London.

Thom, A., 1967. *Megalithic Sites in Britain*, Oxford University Press, Oxford.

Thomas, T., 1983. 'Material Culture of Gatecliff Shelter: Incised Stones', 246–78 in D.H. Thomas *et al.*, *The Archaeology of Monitor Valley 2: Gatecliff Shelter* (*Anthropological Papers of the American Museum of Natural History* 59/1, 1–552).

Thomas, T., 1983. 'The Visual Symbolism of Gatecliff Shelter', 332–52 in D.H. Thomas *et al.*, *The Archaeology of Monitor Valley 2: Gatecliff Shelter* (*Anthropological Papers of the American Museum of Natural History* 59/1, 1–552).

Todd, I.A., 1976. *Çatal Hüyük in Perspective*, Cummings, Menlo Park, California.

Tratman, E.K., 1976. 'A Late Upper Palaeolithic Calculator (?), Gough's Cave, Cheddar, Somerset', *Proceedings of the University of Bristol Spelaeological Society* 14/2, 123–9.

Trinkaus, E., 1982. 'Artificial Cranial Deformation in the Shanidar 1 and 5 Neanderthals', *Current Anthropology* 23/2, 198–9.

Troeng, J., 1993. *Worldwide Chronology of Fifty-three Prehistoric Innovations*, Almqvist & Wiksell, Stockholm.

Turnbull, C.M., 1976. *The Forest People*, Picador, London.

Ucko, P. and A. Rosenfeld, 1967. *Palaeolithic Cave Art*, Weidenfeld and Nicolson, London.

Urry, J., 1989. 'Headhunters and body-snatchers', *Anthropology Today* 5/5, 11–13.

Vandiver, P.B. *et al.*, 1989. 'The Origins of Ceramic Technology at Dolni Vestonice, Czechoslovakia', *Science* 246, 1002–8.

Velo, J., 1984. 'Ochre as Medicine: A Suggestion for the Interpretation of the Archaeological Record', *Current Anthropology* 25/5, 674.

Vencl, S., 1981. 'On Containers in the Palaeolithic and Mesolithic', 309–14 in G. Gramsch (ed.), *Mesolithikum in Europa (Veröffentlichungen des Museums für Ur- und Frühgeschichte Potsdam* 14–15).

Vermeersch, P.M., E. Paulissen and P. Van Peer, 1990. 'Palaeolithic chert exploitation in the limestone stretch of the Egyptian Nile Valley', *African Archaeological Review* 8, 77–102.

Vertes, L., 1965. '"Lunar Calendar" from the Hungarian Upper Paleolithic', *Science* 149, 855–6.

Vishnyatsky, L.B., 1994. '"Running ahead of time" in the development of Palaeolithic industries', *Antiquity* 68, 134–40.

Vlassa, N., 1963. 'Chronology of the Neolithic in Transylvania, in the Light of the Tartaria Settlement's Stratigraphy', *Dacia* N.S. 7, 485–94.

Weinstein-Evron, M. and A. Belfer-Cohen, 1993. 'Natufian Figurines from the New Excavations of the El-Wad Cave, Mt. Carmel, Israel', *Rock Art Research* 10/2, 102–6.

Weisgerber, G., 1987. 'The technological relationship between flint mining and early copper mining', 131–5 in G. de G. Sieveking and M.H. Newcomer (eds.), *The Human Uses of Flint and Chert*, Cambridge University Press, Cambridge.

Wertime, T.A., 1979. 'Pyrotechnology: Man's Fire-Using Crafts', 17–25 in D. Schmandt-Besserat (ed.), *Early Technologies*, Undeno Publications, Malibu.

White, J.P., 1994. 'Site 820 and the Evidence for Early Occupation in Australia', *Quaternary Australasia* 12/2, 21–3.

White, R., 1993. 'Technological and Social Dimensions of "Aurignacian-Age" Body Ornaments across Europe', 277–99 in H. Knecht, A. Pike-Tay and R. White (eds.), *Before Lascaux: The Complex Record of the Early Upper Paleolithic*, CRC Press, Ann Arbor.

Willis, D., 1992. *The Leakey Family: Leaders in the Search for Human Origins*, Facts on File, New York.

Winn, S.M.M., 1981. *Pre-Writing in Southeastern Europe: The Sign System of the Vinca Culture ca. 4000 BC*, Western Publishers, Calgary.

Wreschner, E.E., 1981. 'More on Palaeolithic Ochre', *Current Anthropology* 22/6, 705–6.

Wymer, J.J., 1982. *The Palaeolithic Age*, Croom Helm, London.

Zhang Senshui, 1985. 'The Early Palaeolithic of China', 147–86 in Wu Rukang and J.W. Olsen (eds.), *Palaeoanthropology and Palaeolithic Archaeology in the People's Republic of China*, Academic Press, London.

List of Figures

37 ARTEFACT 1, BILZINGSLEBEN, GERMANY.
38 ARTEFACT 2, BILZINGSLEBEN, GERMANY.
39 ARTEFACT 3, BILZINGSLEBEN, GERMANY.
40 ARTEFACT 4, BILZINGSLEBEN, GERMANY.
41 ADDITIONAL ARTEFACT, BILZINGSLEBEN, GERMANY.

Picture Acknowledgements

PLATES

I, II, XVII, XVIII, XIX, XXI, XXIII, XXIV: ANCIENT ART AND
ARCHITECTURE COLLECTION.
III, IV, V, VI: ROBERT ESTALL.
VII: DRAGOSLAV SREJOVIC.
VIII: ERNEST R. LACHEMAN.
IX: ORIGINALLY PUBLISHED IN M. C. DEWEZ (1974) 'NEW
HYPOTHESES CONCERNING TWO ENGRAVED BONES FROM LA
GROTTE DE REMOUCHAMPS, BELGIUM', *WORLD ARCHAEOLOGY* 5/3
X, XI, XIV, XV: SCIENCE AND SOCIETY PICTURE LIBRARY (SCIENCE
MUSEUM, LONDON).
XII: E. L. MARGETTS.
XIII: PIA BENNIKE.
XVI: BRITISH MUSEUM.
XX, XXII, XXV, XXVII: NOVOSTI (LONDON).
XXVI: MUSEUM OF ARCHAEOLOGY AND ANTHROPOLOGY,
UNIVERSITY OF CAMBRIDGE.
XXVIII: PAUL BAHN.
TOP PHOTOGRAPH ON BACK COVER BY JEAN VERTUT, COURTESY
OF MME. YVONNE VERTUT.

FIGURES

1, 2, 3: ORIGINALLY PUBLISHED IN JAMES MELLAART (1967) *ÇATAL
HÜYÜK: A NEOLITHIC TOWN IN ANATOLIA*, THAMES AND HUDSON.
4: H. MÜLLER-KARPE.
5: M. J. O'KELLY.
6: M. HERITY.
7: ORIGINALLY PUBLISHED IN M. GIMBUTAS (1991) *THE
CIVILISATION OF THE GODDESS*, HARPER SAN FRANCISCO.
8, 9: ORIGINALLY PUBLISHED IN M. GIMBUTAS (1982) *THE
GODDESSES AND GODS OF OLD EUROPE: 6500–3500 BC, MYTHS AND
CULT IMAGES*, THAMES AND HUDSON. ILLUSTRATOR LINDA MOUNT-
WILLIAMS.
10: ORIGINALLY PUBLISHED IN C. RENFREW (1991) 'BEFORE BABEL:
SPECULATIONS ON THE ORIGINS OF LINGUISTIC

DIVERSITY', *CAMBRIDGE ARCHAEOLOGICAL JOURNAL* 1/1.
11: ORIGINALLY PUBLISHED IN D. SCHMANDT-BESSERAT (1992)
BEFORE WRITING, VOLUME ONE: FROM COUNTING TO CUNEIFORM,
UNIVERSITY OF TEXAS PRESS.
12: N. VLASSA.
13: ORIGINALLY PUBLISHED IN S. M. M. WINN (1981) *PRE-WRITING IN
SOUTHEASTERN EUROPE: THE SIGN SYSTEM OF THE VINCA CULTURE*,
WESTERN PUBLISHERS.
15: HARALD HAARMANN.
16, 17, 18: ORIGINALLY PUBLISHED IN A. FORBES AND T. R.
CROWDER (1979) 'THE PROBLEM OF FRANCO-CANTABRIAN
ABSTRACT SIGNS: AGENDA FOR A NEW APPROACH', *WORLD
ARCHAEOLOGY* 10/3.
19: ORIGINALLY PUBLISHED IN F. D'ERRICO (1995) 'A NEW MODEL
AND ITS IMPLICATIONS FOR THE ORIGIN OF WRITING: THE LA
MARCHE ANTLER REVISITED', *CAMBRIDGE ARCHAEOLOGICAL
JOURNAL* 5/2.
20: JOHN GOWLETT.
21: ORIGINALLY PUBLISHED IN M. C. DEWEZ (1974) 'NEW
HYPOTHESES CONCERNING TWO ENGRAVED BONES FROM LA
GROTTE DE REMOUCHAMPS, BELGIUM', *WORLD ARCHAEOLOGY* 5/3.
22: ORIGINALLY PUBLISHED IN E. K. TRATMAN (1976) 'A LATE
UPPER PALAEOLITHIC CALCULATOR (?), GOUGH'S CAVE, CHEDDAR,
SOMERSET', *PROCEEDINGS OF THE UNIVERSITY OF BRISTOL
SPELAEOLOGICAL SOCIETY* 14/2. COURTESY OF THE UNIVERSITY OF
BRISTOL SPELAEOLOGICAL SOCIETY.
23: L. VÉRTES.
24: ORIGINALLY PUBLISHED IN L. LIEBENBERG (1990) *THE ART OF
TRACKING: THE ORIGIN OF SCIENCE*, DAVID PHILIP.
26: G. HENRIKSEN.
27: ORIGINALLY PUBLISHED IN R. SHEPHERD (1980) *PREHISTORIC
MINING AND ALLIED INDUSTRIES*, ACADEMIC PRESS. BY PERMISSION
OF THE PUBLISHER ACADEMIC PRESS.
28: ORIGINALLY PUBLISHED IN R. HOLGATE (1991) *PREHISTORIC
FLINT MINES*, SHIRE PUBLICATIONS.
29: ORIGINALLY PUBLISHED IN C. M. AIKENS AND T. HIGUCHI (1982)
PREHISTORY OF JAPAN, ACADEMIC PRESS. BY PERMISSION OF THE
PUBLISHER ACADEMIC PRESS.
30, 31: ORIGINALLY PUBLISHED IN B. BOYD AND J. COOK (1993) 'A
RECONSIDERATION OF THE "AIN SAKHRI" FIGURINE',
PROCEEDINGS OF THE PREHISTORIC SOCIETY 59. ILLUSTRATOR PHIL
DEAN. REPRODUCTION BY KIND PERMISSION OF THE TRUSTEES OF
THE BRITISH MUSEUM.
32, 41: ROBERT BEDNARIK.
33, 34: DIRK HUYGE AND ROBERT BEDNARIK.
35: ORIGINALLY PUBLISHED IN T. MALINOWSKI (1981)
'ARCHAEOLOGY AND MUSICAL INSTRUMENTS IN POLAND', *WORLD
ARCHAEOLOGY* 12/3.
36: U. FISCHER.
37, 38, 39, 40: D. AND U. MANIA, ROBERT BEDNARIK.

Index

Note: Page numbers for illustrations are in *italics*.